# Essential Algebra

## A Self-Teaching Guide

Tim Hill

Questing Vole Press

*Essential Algebra: A Self-Teaching Guide*
by Tim Hill

Editor: Kevin Debenjak
Proofreader: Diane Yee
Compositor: Kim Frees
Cover: Questing Vole Press

Second Edition

# Contents

# 1         A Few Basics

Before jumping into algebra, you need to know a few basic facts about arithmetic, notation, and real numbers. Feel free to skip or skim this chapter if you're already familiar with the material.

## Algebraic Notation

Arithmetic involves calculations with particular numbers:

$$(8 + 3)(8 - 3) = 11 \cdot 5 = 55$$

whereas algebra states universal facts:

$$(a + b)(a - b) = a^2 - b^2$$

Specific arithmetic calculations are of only limited interest, but general algebraic results are valuable.

Algebraic notation uses **variables** to represent unspecified numbers. Letters at the beginning of the alphabet ($a$, $b$, $c$,...) typically represent constants, and letters at the end the alphabet ($x$, $y$, $z$,...) typically represent variable or unknown quantities. Greek letters ($\alpha$, $\beta$, $\gamma$,...) also appear in precalculus, mostly in trigonometry. These rules aren't strict, and the statement

$$(a + b)(a - b) = a^2 - b^2$$

means exactly the same as

$$(x + y)(x - y) = x^2 - y^2$$

1

## Grouping Expressions

A quantity surrounded by parentheses (), brackets [], or braces {} is treated as a single number. If parenthetic expressions are nested, evaluate the innermost parentheses first and then work outward. In the following calculation, symbols are removed from the inside out, changing signs throughout the grouped expression when the grouping symbol is preceded by a minus sign.

$$a - \left[ 4a - \left( 3a + b \right) \right] = a - \left[ 4a - 3a - b \right]$$
$$= a - 4a + 3a + b$$
$$= b$$

Grouping symbols also prevents ambiguity and misunderstanding:

$$1 - (a + b) \text{ is not the same as } 1 - a + b$$

$$\frac{1}{2}(a + b) \text{ is not the same as } \frac{1}{2}a + b$$

$$1 / (a + b) = \frac{1}{a + b}$$

$$1 / a + b = \frac{1}{a} + b$$

## Order of Algebraic Operations

The expression $2 + 3 \cdot 7$ contains no parentheses that indicate which operation should be performed first. By convention, however, multiplication and division are performed before addition and subtraction unless parentheses indicate otherwise. Thus $2 + 3 \cdot 7$ means $2 + (3 \cdot 7)$, which equals 23—not $(2 + 3) \cdot 7 = 35$. Similarly, $a + (bc)$ means $a + bc$. Omitting unnecessary parentheses makes expressions cleaner and easier to read.

## Don't Divide by Zero

Any operation that involves division by zero is meaningless. To see why, observe that $a/0 = b$ and $a = 0 \cdot b$ would both mean the same thing if division by zero were valid. But if $a \neq 0$, then $a = 0 \cdot b$ isn't true for any $b$; and if $a = 0$, then $a = 0 \cdot b$ is true for every $b$. This result shows that $a/0$ can't be assigned a definite value and therefore has no real meaning.

# The Real Numbers

The basic numbers used in algebra are the **real numbers**. Defining the real number system formally requires advanced mathematics, but all that's required in algebra is a solid grounding and an intuitive grasp.

The real number system contains different types of numbers. The **positive integers** (or **natural numbers**) are the numbers

$$1, 2, 3, 4, 5, \ldots$$

The **integers** are the numbers

$$\ldots, -3, -2, -1, 0, 1, 2, 3, \ldots$$

The **rational numbers** are numbers of the form

$$\frac{m}{n}$$

where $m$ and $n$ are integers and $n \neq 0$. Examples of rational numbers are

$$\frac{1}{2}, -\frac{7}{5}, 0, 3, 6.8622, -5, 4\frac{2}{3}$$

Note that you can add, subtract, multiply, and divide rational numbers and stay within the system of rational numbers. Rational numbers suffice for all physical measurements (weight, length, volume, and so on) of any accuracy. Algebra, geometry, and calculus, however, require a richer system of numbers that includes irrational numbers. A real number that isn't a rational number is an **irrational number**, for example,

$$\sqrt{2}, \sqrt{3}, \sqrt{5}, \pi, \pi/5, -\sqrt{3}$$

(Recall that $\sqrt{a}$ means the *positive* square root of any positive number $a$. For example, $\sqrt{4}$ is equal to 2 and not –2, even though $(-2)^2 = 4$.)

The types of real numbers form increasingly inclusive subsets (Figure 1.1).

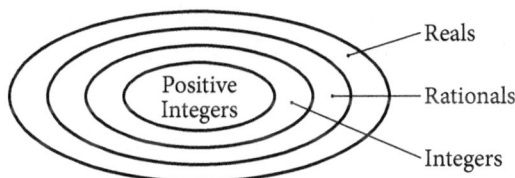

**Figure 1.1** Types of real numbers

You can also separate the real numbers into the rationals and the irrationals (Figure 1.2).

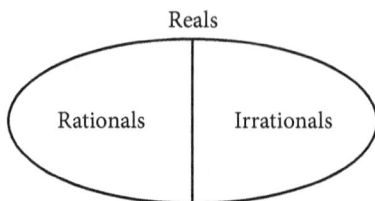

**Figure 1.2** Rational and irrational numbers make up the real numbers

## The Real Line

The best way to picture the real number system is by using the **real line** (also called the **number line**). Imagine a horizontal straight line, extending endlessly in both directions. Choose an arbitrary point, called the **origin** or **zero point**, and label it 0. Pick another arbitrary point to the right of 0 and label it 1. The distance between these two points (the **unit distance**) is the measuring scale that associates every real number with a point on the line (Figure 1.3).

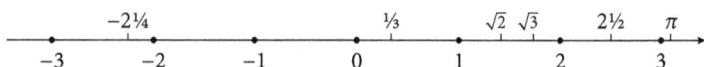

**Figure 1.3** The real line

You can plot irrational numbers on the line by using their decimal expansions: $\sqrt{2} = 1.414\ldots$, $\sqrt{3} = 1.732\ldots$, and $\pi = 3.14159\ldots$. (The decimal expansion of an irrational number never repeats or terminates, unlike a rational number.)

The line has no gaps and every distance can be represented by a point on the line. The real numbers can be described as all those numbers that correspond to all points on the real line.

## Commutative, Associative, and Distributive Laws

The **commutative law** states that order doesn't matter in addition and multiplication:

$$a + b = b + a \quad \text{and} \quad ab = ba$$

The **associative law** states that grouping doesn't matter in addition and multiplication:

$$(a + b) + c = a + (b + c) \quad \text{and} \quad (ab)c = a(bc)$$

The **distributive law** connects addition and multiplication, converting a product with a sum into a sum of two products:

$$a(b + c) = ab + ac$$

Because multiplication is commutative, this law can be rewritten as

$$(a + b)c = ac + bc$$

Depending on the context, the distributive law can be used in either direction: to transform $a(b + c)$ into $ab + ac$, or to transform $ab + ac$ into $a(b + c)$. This law is often used for quick calculations, for example,

$$17 \cdot 11 = 17(10 + 1) = 17 \cdot 10 + 17 \cdot 1 = 170 + 17 = 187$$

and

$$23 \cdot 19 = 23(20 - 1) = 460 - 23 = 437$$

When written in reverse order, $ab + ac = a(b + c)$, the distributive law justifies the factoring of expressions. The two terms on the left have $a$ as a common factor, which is factored out on the right. Factoring is a crucial tool in simplifying algebraic expressions, for example,

$$12abc - 6ac + 138ad = 6a(2bc - c + 23d)$$

and

$$22a(a-b) - 40(a-b)^2 = 2(a-b) \times [11a - 20(a-b)]$$

$$= 2(a-b) \times [11a - 20a + 20b]$$

$$= 2(a-b) \times [20b - 9a]$$

The following identities arise from the distributive law:

$$(a + b)^2 = a^2 + 2ab + b^2$$

$$(a - b)^2 = a^2 - 2ab + b^2$$

$$(a + b)(a - b) = a^2 - b^2$$

# Fractions

Fractions are added, subtracted, multiplied, and divided by the following rules:

$$\frac{a}{b} + \frac{c}{d} = \frac{ad+bc}{bd}$$ To get the numerator, cross-multiply and add.

$$\frac{a}{b} - \frac{c}{d} = \frac{ad-bc}{bd}$$ To get the numerator, cross-multiply and subtract.

$$\frac{a}{b} \cdot \frac{c}{d} = \frac{ac}{bd}$$

$$\frac{\frac{a}{b}}{\frac{c}{d}} = \frac{a}{b} \cdot \frac{d}{c}$$ To divide, invert the denominator and multiply.

Here are some examples:

$$\frac{1}{a+b} + \frac{1}{a-b} = \frac{(a-b)+(a+b)}{(a+b)(a-b)} = \frac{2a}{a^2-b^2}$$

$$\frac{3a}{2} - \frac{4a}{3} = \frac{9a-8a}{6} = \frac{a}{6}$$

$$\frac{1}{a(b+c)} \cdot \frac{a}{(b-c)} = \frac{1}{b^2-c^2}$$

$$\frac{\frac{18abc}{2d}}{\frac{3}{ad}} = \frac{18abc}{2d} \cdot \frac{ad}{3} = \frac{18a^2bcd}{6d} = 3a^2bc$$

# Problems

1. Simplify the following expressions by removing parentheses, brackets, and braces.

   $(3a - b) - [2a - (a + b)]$

   $[(a + 3b) - a] - [a - (a - 3b)]$

   $a - \{2a - [b - (3a - 2b)]\}$

2. Simplify the following expression.

   $$\frac{2}{5} \cdot \frac{m+3}{7} + \frac{1}{2}$$

3. Factor the following expressions.

   $12x - 18y + 30 \qquad 8x^2 - 12x^3y - 28x^4z \qquad 9abc + 3a^2b^2c^2$

4. Simplify the following expressions.

   $$\frac{a}{b} - \frac{b}{a} \qquad \frac{1}{1 + \dfrac{1}{x-1}} \qquad \frac{1}{x-y}\left(\frac{x}{y} - \frac{y}{x}\right) \qquad \frac{(x+a)^2 - x^2}{a}$$

5. Simplify the following expression.

   $$\frac{x-2}{y} \left/ \frac{z}{x+2} \right.$$

# 2

# Exponents

This chapter covers the algebra of exponentiation in preparation for dealing with polynomials, quadratic equations, and special classes of functions.

## Integer Exponents

You've already used exponent notation for squares, cubes, and fourth powers: $a^2 = a \cdot a$, $a^3 = a \cdot a \cdot a$, and $a^4 = a \cdot a \cdot a \cdot a$. In the same way, $a^n$ is defined by $a^n = a \cdot a \cdots a$ ($n$ factors) for any positive integer $n$. The rules of exponents are:

| Rule | Example |
|------|---------|
| $a^m a^n = a^{m+n}$ | $a^2 a^3 = \underbrace{(a \cdot a)}_{2}\underbrace{(a \cdot a \cdot a)}_{3} = \underbrace{a \cdot a \cdot a \cdot a \cdot a}_{5 \text{ factors}} = a^5$ |
| $\dfrac{a^m}{a^n} = a^{m-n}$ | $\dfrac{a^5}{a^3} = \dfrac{a \cdot a \cdot a \cdot a \cdot a}{a \cdot a \cdot a} = \dfrac{a \cdot a \cdot a}{a \cdot a \cdot a} \cdot \dfrac{a \cdot a}{1} = a \cdot a = a^2$ |
| $(a^m)^n = a^{mn}$ | $(a^3)^2 = (a \cdot a \cdot a)(a \cdot a \cdot a) = a \cdot a \cdot a \cdot a \cdot a \cdot a = a^6$ |
| $(ab)^n = a^n b^n$ | $(ab)^3 = (a \cdot b)(a \cdot b)(a \cdot b) = a \cdot a \cdot a \cdot b \cdot b \cdot b = a^3 b^3$ |
| $\left(\dfrac{a}{b}\right)^n = \dfrac{a^n}{b^n}$ | $\left(\dfrac{a}{b}\right)^4 = \left(\dfrac{a}{b}\right)\cdot\left(\dfrac{a}{b}\right)\cdot\left(\dfrac{a}{b}\right)\cdot\left(\dfrac{a}{b}\right) = \dfrac{a \cdot a \cdot a \cdot a}{b \cdot b \cdot b \cdot b} = \dfrac{a^4}{b^4}$ |

9

If $a \neq 0$, then the rule $a^m a^n = a^{m+n}$ suggests a logical way to define $a^0$. Because we want the equation $a^0 a^n = a^{0+n} = a^n$ to be true, $a^0$ should leave $a^n$ unchanged by multiplication. Therefore, $a^0$ is defined by

$$a^0 = 1$$

Similarly, for the equation $a^n a^{-n} = a^{n-n} = a^0 = 1$ to hold, $a^{-n}$ must be the reciprocal of $a^n$. Therefore, $a^{-n}$ is defined by

$$a^{-n} = \frac{1}{a^n}$$

With these two definitions, the rules of exponents hold for all integer exponents, positive, negative, or zero.

You can move any factor $a^n$ from the numerator to the denominator or vice versa by changing the sign of the exponent. The following examples show how to rewrite expressions to have all positive exponents.

$$a^5 b^{-6} = a^5 \cdot \frac{1}{b^6} = \frac{a^5}{b^6}$$

$$\frac{1}{a^{-4}} = \frac{1}{\dfrac{1}{a^4}} = a^4$$

$$\frac{a^{-2}}{b^3} = \frac{\dfrac{1}{a^2}}{b^3} = \frac{1}{a^2 b^3}$$

## Roots

You've already seen $\sqrt{2}$, $\sqrt{3}$, and other square roots. This section generalizes the ideas of roots.

If $n$ is a positive integer and $x^n = a$, then $x$ is called an **nth root** of $a$. In particular, $x$ is called a **square root** of $a$ if $x^2 = a$, and a **cube root** of $a$ if $x^3 = a$. For example,

- $2^2 = 4$ and $(-2)^2 = 4$, so 2 and $-2$ are both square roots of 4.

- $2^3 = 8$, so 2 is a cube root of 8.

- $(-2)^3 = -8$, so $-2$ is a cube root of $-8$.

- $3^4 = 81$ and $(-3)^4 = 81$, so $3$ and $-3$ are both fourth roots of 81.

- $2^5 = 32$, so 2 is a fifth root of 32.

## Radicals

Because the square of a positive or negative real number is positive, negative numbers have no real square roots. Each positive number $a$, however, has two square roots, numerically equal but opposite in sign, with the positive root denoted by $\sqrt{a}$ and the negative root by $-\sqrt{a}$. Similarly, if $a$ is positive, then it has a single positive $n$th root denoted by $\sqrt[n]{a}$. The symbol $\sqrt{\phantom{x}}$ is called the **radical sign**, and the number $n$ is called the **index** of the root. As shown previously, in the case of the square root ($n = 2$) the index is omitted. For example,

$$\sqrt{36} = 6 \quad \sqrt[3]{-27} = -3 \quad \sqrt[4]{16} = 2 \quad \sqrt[5]{-1} = -1 \quad \sqrt[6]{-1} \text{ does not exist}$$

The basic facts are:

- If $a$ is positive and $n$ is even or odd, then $\sqrt[n]{a}$ is positive.

- If $a$ is negative and $n$ is odd, then $\sqrt[n]{a}$ is negative.

- If $a$ is negative and $n$ is even, then $\sqrt[n]{a}$ does not exist.

The rules of radicals are:

| Rule | Example |
|---|---|
| $(\sqrt[n]{a})^n = a$ | $(\sqrt{2})^2 = 2$ |
| $\sqrt[n]{a^n} = a$ | $\sqrt[3]{125} = \sqrt[3]{5^3} = 5$ |
| $\sqrt[n]{ab} = \sqrt[n]{a} \cdot \sqrt[n]{b}$ | $\sqrt{50} = \sqrt{25 \cdot 2} = \sqrt{25} \cdot \sqrt{2} = 5\sqrt{2}$ |
| $\sqrt[n]{\dfrac{a}{b}} = \dfrac{\sqrt[n]{a}}{\sqrt[n]{b}}$ | $\sqrt{\dfrac{4}{9}} = \dfrac{\sqrt{4}}{\sqrt{9}} = \dfrac{2}{3}$ |
| $\sqrt[m]{\sqrt[n]{a}} = \sqrt[mn]{a}$ | $\sqrt[6]{a^2} = \sqrt[3]{\sqrt{a^2}} = \sqrt[3]{a}$ |

## Rationalizing the Denominator

In calculations, it's often convenient to **rationalize the denominator,** that is, remove square roots from the denominators of fractional expressions. The following calculations show two ways to do this.

$$\frac{2}{\sqrt{2}} = \frac{2}{\sqrt{2}} \cdot \frac{\sqrt{2}}{\sqrt{2}} = \frac{2\sqrt{2}}{2} = \sqrt{2}$$

$$\frac{2}{\sqrt{3}+1} = \frac{2}{\sqrt{3}+1} \cdot \frac{\sqrt{3}-1}{\sqrt{3}-1} = \frac{2(\sqrt{3}-1)}{(\sqrt{3})^2 - 1^2} = \frac{2(\sqrt{3}-1)}{3-1} = \sqrt{3}-1$$

## Rational Exponents

This section defines exponentiation by rational numbers in such a way that the rules for integer exponents (page 9) remain valid. Rational numbers take the form $m/n$, where $m$ and $n$ are integers and $n \neq 0$.

Because we want the equation $(a^{1/2})^2 = a^{(1/2) \cdot 2} = a^1 = a$ to be true, $a^{1/2}$ is defined as

$$a^{1/2} = \sqrt{a}$$

Similarly,

$$a^{1/3} = \sqrt[3]{a}$$

In general,

$$a^{1/n} = \sqrt[n]{a}$$

for any positive integer $n$. For example,

$$9^{1/2} = \sqrt{9} = 3 \qquad\qquad 64^{1/3} = \sqrt[3]{64} = 4$$

$$(-27)^{1/3} = \sqrt[3]{-27} = -3 \qquad\qquad 16^{1/4} = \sqrt[4]{16} = 2$$

Now that the expression $a^{1/n}$ makes sense, we can generalize to $a^{m/n}$, where $m/n$ is a rational number. Assume that any fraction used as an exponent is written in **lowest terms,** that is, in the form $m/n$ where $n$ is a positive integer, $m$ is an integer (positive, negative, or zero), and $m$ and $n$ have no common factors greater than 1. We want $(a^{m/n})^n = a^m$ to hold, so that $a^{m/n}$ is the $n$th root of $a^m$. Therefore, $a^{m/n}$ is defined as

$$a^{m/n} = \sqrt[n]{a^m}$$

For example,

$$4^{3/2} = \sqrt{4^3} = \sqrt{64} = 8$$

$$8^{2/3} = \sqrt[3]{8^2} = \sqrt[3]{64} = 4$$

$$a^{2/7} \cdot a^{5/2} = a^{4/14} \cdot a^{35/14} = a^{39/14} = \sqrt[14]{a^{39}}$$

$$\frac{a^{2/3}}{a^{3/5}} = a^{(2/3-3/5)} = a^{1/15} = \sqrt[15]{a}$$

$$\frac{\sqrt[3]{a^4}}{\sqrt[4]{a^3}} = \frac{a^{4/3}}{a^{3/4}} = a^{(4/3-3/4)} = a^{7/12} = \sqrt[12]{a^7}$$

It's sometimes handy to use the fact

$$a^{m/n} = \sqrt[n]{a^m} = (\sqrt[n]{a})^m$$

in calculations. For example, $8^{2/3}$ is easy to evaluate either as

$$8^{2/3} = \sqrt[3]{8^2} = \sqrt[3]{64} = 4$$

or as

$$8^{2/3} = (\sqrt[3]{8})^2 = 2^2 = 4$$

However, $32^{3/5}$ is hard to evaluate as

$$32^{3/5} = \sqrt[5]{32^3}$$

but easy to evaluate as

$$32^{3/5} = (\sqrt[5]{32})^3 = 2^3 = 8$$

## Problems

1. Simplify the following expressions.

$$x^5(x^2)^3 \qquad y^4(y^2(y^5)^2)^3 \qquad t^4(t^3(t^{-2})^5)^4$$

$$\frac{(x^{-2})^3 y^8}{x^{-5}(y^4)^{-3}} \qquad \frac{(x^2 y^4)^3}{(x^5 y^2)^{-4}} \qquad \left(\frac{(x^2 y^{-5})^{-4}}{(x^5 y^{-2})^{-3}}\right)^2$$

2. Find integers $m$ and $n$ such that $2^m \cdot 5^n = 16000$.

3. Write $8^{1000}/2^5$ as a power of 2.

4. Simplify the following expressions.

$$(a^{n-4}b^4)(ab^{n-1})^4 \quad (4a^3b^{-4})(3a^{-1}b^5) \quad \frac{x^{14}y^5}{x^4y^{-5}}$$

$$a^2b^2(a^{-2}+b^{-2}) \quad (x+y)(x^{-1}+y^{-1}) \quad \left(\frac{a^2b}{c}\right)^4\left(\frac{a}{b^2c^3}\right)^2\left(\frac{c^2}{a^2}\right)^5$$

5. Simplify the following expressions.

$$\sqrt{144} \qquad \sqrt{0.64} \qquad \sqrt{\frac{225}{400}} \qquad \sqrt[3]{-\frac{1}{27}}$$

$$\sqrt[3]{-1000} \qquad \sqrt{125} \qquad \sqrt[4]{625} \qquad \sqrt{18}$$

$$\sqrt{12} \qquad \sqrt{2}+\sqrt{8} \qquad \sqrt{3}+\sqrt[4]{9} \qquad \sqrt[3]{54}+\sqrt[3]{250}$$

$$\sqrt[10]{32a^5} \qquad \sqrt{a^2b^4} \qquad \sqrt[4]{a^5} \qquad \sqrt{1-\left(\frac{\sqrt{3}}{2}\right)^2}$$

6. Rationalize the denominator of the following expressions.

$$\frac{30}{\sqrt{6}} \quad \frac{\sqrt{6}+2}{\sqrt{6}-2} \quad \frac{2}{\sqrt{7}+\sqrt{5}}$$

7. Compute the following values.

$$25^{3/2} \quad 32^{3/5} \quad 32^{-4/5} \quad (-8)^{7/3}$$

8. Simplify the following expressions.

$$(25a^6b^{-2})^{1/2} \qquad (2a^{1/2}b^{1/4})^4 \qquad \sqrt[5]{a^2b}\cdot\sqrt[5]{a^3b^4}$$

$$(a^{1/2}+b^{1/2})(a^{1/2}-b^{1/2}) \quad \left\{a^{2/3}\cdot\left[\left(\frac{a^{2/3}}{a^{1/4}}\right)^6\right]^{1/3}\right\}^2 \quad \left(\frac{27b^2c^5}{64a^6b^{-4}c^{-1}}\right)^{1/3}$$

# 3

# Polynomials

A **polynomial**, or more precisely, a **polynomial in** $x$, is an algebraic expression that's built up from the variable $x$ and any constants by means of addition, subtraction, and multiplication alone. Recall from "Algebraic Notation" in Chapter 1 that later letters $(x, y, z, \ldots)$ are used for variables and earlier letters $(a, b, c, \ldots)$ are used for constants. Specific numbers like 5, $\sqrt{2}$, and $\pi$ are also constants.

Examples of polynomials are

$$1 \qquad x \qquad x + 1 \qquad 2x^2 \qquad 2x^2 + x + 1$$

and

$$2x^5 + 8x^4 - 3x^3 + 622x^2 - 6x + 19$$

In the preceding example, the constants 2, 8, −3, 622, −6, and 19 are called **coefficients** of the polynomial, and the exponent of the highest power of $x$—here, 5—is called its **degree**. Certain polynomials are classified by their degree:

- **Constant polynomial** (degree 0): $a$ $(a \neq 0)$.

- **Linear polynomial** (degree 1): $ax + b$ $(a \neq 0)$.

- **Quadratic polynomial** (degree 2): $ax^2 + bx + c$ $(a \neq 0)$.

- **Cubic polynomial** (degree 3): $ax^3 + bx^2 + cx + d$ $(a \neq 0)$.

The constant polynomial 0 isn't assigned a degree.

Polynomials are added and subtracted by combining terms that have the same powers of $x$. For example,

$$(5x^3 - 5x^2 + 10x - 5) + (8x^2 + 2x + 16) = 5x^3 + 3x^2 + 12x + 11$$

$$(7x^4 + 2x^2 - 3) - (6x^3 + 2x^2 + 8x + 2) = 7x^4 - 6x^3 - 8x - 5$$

Polynomials are multiplied like any other sums. To calculate the product of two polynomial factors, multiply the second factor by each term in the first factor, simplify by using the exponent rule $x^m \cdot x^n = x^{m+n}$, and then collect terms that have the same powers of $x$. For example,

$$(x^2 - 2x + 1) \cdot (3x^3 + 4x^2 + 5x)$$

$$= 3x^5 + 4x^4 + 5x^3 - 6x^4 - 8x^3 - 10x^2 + 3x^3 + 4x^2 + 5x$$

$$= 3x^5 - 2x^4 - 6x^2 + 5x$$

Note that the degree of the product of two nonzero polynomials equals the sum of their individual degrees.

To divide polynomials, see Chapter 14.

## Problems

1. Simplify the following expressions.

$$(x^7 - 3x^5 + 4x^2 - 9) + (2x^6 - 5x^5 - 2x^4 + x^3 - 2x^2 + x + 1)$$

$$(3x^5 + x^4 - 2x^3 + 5x^2 - 11x + 2) - (x^4 + 5x^2 + 2)$$

$$(2x^3 + 3x^2 - 4)(3x^2 - 2x - 9)$$

$$(x^5 - 2x^3 + 3)(2x^2 - 8x + 4)$$

$$(x - 1)(x^3 + x^2 + x + 1)$$

# 4

# Factoring

Factoring a polynomial expresses it as a product of polynomials of lower degrees. Factoring is useful for solving certain types of equations.

The simplest type of factoring removes a common polynomial factor (which should always be done first). For example,

$$x^4 + x^2 = x^2(x^2 + 1)$$

$$\tfrac{1}{2}x^4 - \tfrac{1}{4}x^2 = \tfrac{1}{4}x^2(2x^2 - 1)$$

$$3x^5 + 33x^4 - 12x^3 = 3x^3(x^2 + 11x - 4)$$

$$2(x - 5)^3 - 8(x - 5)^2 = 2(x - 5)^2[(x - 5) - 4] = 2(x - 5)^2(x - 9)$$

Certain products occur so often that you should memorize them:

(1) $\quad (x + a)(x - a) = x^2 - a^2$

(2) $\quad (x + a)(x + a) = (x + a)^2 = x^2 + 2ax + a^2$

(3) $\quad (x - a)(x - a) = (x - a)^2 = x^2 - 2ax + a^2$

(4) $\quad (x + a)(x + b) = x^2 + (a + b)x + ab$

(5) $\quad (ax + b)(cx + d) = acx^2 + (ad + bc)x + bd$

Reading from left to right, each of the preceding products is an **expansion formula**. Reading from right to left, each product is a **factoring formula**.

The following examples illustrate (1), (2), and (3).

$$x^2 - 16 = (x + 4)(x - 4)$$

$$4x^2 - 25 = (2x)^2 - 5^2 = (2x + 5)(2x - 5)$$

$$x^2 + 6x + 9 = (x + 3)^2$$

$$x^2 - 10x + 25 = (x - 5)^2$$

The trick in using (4) to factor a polynomial of the form $x^2 + px + q$ is to test various pairs of numbers $a$ and $b$ whose product is $q$ in the hopes that one of the pairs sums to $p$. For example,

$$x^2 + x - 6 = (x + 3)(x - 2)$$

$$x^2 + 8x + 15 = (x + 3)(x + 5)$$

$$x^2 - 9x + 14 = (x - 2)(x - 7)$$

Using (5) is similar but more laborious because patient trial and error of various combinations of $a$, $b$, $c$, and $d$ is needed. For example,

$$2x^2 + 5x - 3 = (2x - 1)(x + 3)$$

$$3x^2 + x - 2 = (x + 1)(3x - 2)$$

$$12x^2 + 7x - 10 = (3x - 2)(4x + 5)$$

$$8x^2 + 10x - 12 = (4x - 3)(2x + 4)$$

## Problems

1. Factor the following expressions.

   $x^2 - x - 6$     $x^2 - 4x + 4$     $x^3 + 12x^2 + 36x$

   $x^4 - 16$     $x^3 - 3x^2 - 4x$     $4x^2 + 2x - 12$

   $10x^2 - 16x - 8$

2. Verify the formula $(x - a)(x^2 + ax + a^2) = x^3 - a^3$ and use it to factor $x^3 - 27$ and $8x^3 - 125$.

3. Verify the formula $(x + a)(x^2 - ax + a^2) = x^3 + a^3$ and use it to factor $x^3 + 64$ and $27x^3 + 8$.

4. Solve the following equations by factoring.

   $x^2 + 3x - 28 = 0$

   $x^2 - 8x - 33 = 0$

   $2x^2 + x - 15 = 0$

   $6x^2 - 5x - 21 = 0$

# 5 Linear & Quadratic Equations

A **linear equation** is an equation like $2x - 4 = 0$ or $3x + 21 = 0$. The general form is

$$ax + b = 0 \qquad (a \neq 0)$$

Solving for the unknown $x$ is simple: subtract the constant $b$ from both sides ($ax = -b$) and then divide both sides by the coefficient $a$, so

$$x = -\frac{b}{a}$$

For example, the solution to $2x - 4 = 0$ is $x = 2$, and the solution to $3x + 21 = 0$ is $x = -7$.

A **quadratic equation** is an equation like $x^2 + x - 12$. The general form is

$$ax^2 + bx + c = 0 \qquad (a \neq 0)$$

Some quadratic equations can be solved by factoring (Chapter 4). For example, the factored form of $x^2 + x - 12 = 0$ is $(x + 4)(x - 3) = 0$. To get the **roots** (solutions), set each factor to zero: $x + 4 = 0$, so $x = -4$; and $x - 3 = 0$, so $x = 3$. The roots are $-4$ and $3$.

Finding roots in this way rests on the fact that the product of two numbers equals zero if and only if one of the factors is zero. In the preceding example, $x^2 + x - 12 = 0$, or equivalently $(x + 4)(x - 3) = 0$, is satisfied if and only if $x + 4 = 0$ or $x - 3 = 0$, and each of these equations yield one of the roots.

If factoring fails, use the general formula—called the **quadratic formula**—to get the roots:

$$x = \frac{-b \pm \sqrt{b^2 - 4ac}}{2a}$$

Applying this formula to the preceding example yields, as before,

$$x = \frac{-1 \pm \sqrt{1 + 48}}{2} = \frac{-1 \pm \sqrt{49}}{2} = \frac{-1 \pm 7}{2} = -4 \text{ or } 3$$

Before deriving the quadratic formula, note that the square of $x + t$ is

$$(x + t)^2 = x^2 + 2tx + t^2$$

The right side of this formula is a perfect square because its constant term ($t^2$) is the square of half the coefficient of $x$ ($2t$), demonstrating a technique called **completing the square**, which states that

$$x^2 + bx = \left(x + \frac{b}{2}\right)^2 - \left(\frac{b}{2}\right)^2$$

The derivation of the quadratic formula is

| | |
|---|---|
| $ax^2 + bx + c = 0$ | Given $(a \neq 0)$. |
| $x^2 + \dfrac{b}{a}x + \dfrac{c}{a} = 0$ | Divide by $a$. |
| $x^2 + \dfrac{b}{a}x = -\dfrac{c}{a}$ | Move the constant to the right. |
| $x^2 + \dfrac{b}{a}x + \dfrac{b^2}{4a^2} = -\dfrac{c}{a} + \dfrac{b^2}{4a^2}$ | Complete the square. |
| $\left(x + \dfrac{b}{2a}\right)^2 = \dfrac{b^2 - 4ac}{4a^2}$ | Factor and simplify. |
| $x + \dfrac{b}{2a} = \dfrac{\pm\sqrt{b^2 - 4ac}}{2a}$ | Take square roots. |
| $x = \dfrac{-b \pm \sqrt{b^2 - 4ac}}{2a}$ | Solve for $x$. |

In the preceding derivation, we completed the square on the left by adding $(b/2a)^2 = b^2/4a^2$ (the square of half the coefficient of $x$), and adding the same number on the right.

For example, applying the quadratic formula to $x^2 - 6x + 6 = 0$ yields $a = 1$, $b = -6$, and $c = 6$, and the roots are

$$x = \frac{6 \pm \sqrt{36-24}}{2} = \frac{6 \pm 2\sqrt{3}}{2} = 3 \pm \sqrt{3}$$

Applying the quadratic formula to $x^2 + 4x + 4 = 0$ yields $a = 1$, $b = 4$, and $c = 4$, and the roots are

$$x = \frac{-4 \pm \sqrt{16-16}}{2} = \frac{-4 \pm 0}{2} = -2, -2$$

Applying the quadratic formula to $x^2 - 2x + 2 = 0$ yields $a = 1$, $b = -2$, and $c = 2$, and the roots are

$$x = \frac{2 \pm \sqrt{4-8}}{2} = \frac{2 \pm 2\sqrt{-1}}{2} = 1 \pm \sqrt{-1}$$

The three preceding examples show the three possibilities for the roots of a quadratic equation:

- Two real roots

- One real root

- Two imaginary roots (no real roots)

Imaginary roots involve the square roots of negative numbers, which aren't real numbers (and are beyond the scope of this book). The geometric meaning of the solutions to quadratic functions is covered in Chapter 12.

## Problems

1. Solve the following equations by using the quadratic formula.

$$5x^2 - 9x + 3 = 0 \quad 3x^2 + 7x + 3 = 0$$

$$17x^2 - 6x + 1 = 0 \quad x^2 + x + 1 = 0$$

# 6

# Inequalities & Absolute Values

The left-to-right linear arrangement of real numbers as points on the real line makes it easy to visualize the algebra of inequalities.

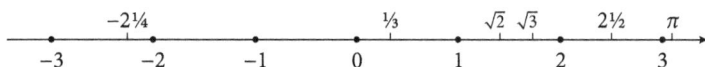

The inequality $a < b$ (read "$a$ is less than $b$") means that the point $a$ lies to the left of the point $b$, and the equivalent inequality $b > a$ (read "$b$ is greater than $a$") means that $b$ lies to the right of $a$. A number $a$ is positive or negative depending on whether $a > 0$ or $a < 0$. The main rules for inequalities are:

- If $a > 0$ and $b < c$, then $ab < ac$.

- If $a < 0$ and $b < c$, then $ab > ac$.

- If $a < b$, then $a + c < b + c$ for all $c$.

The first two rules say that inequality is preserved on multiplication by a positive number, and reversed on multiplication by a negative number. The third rules says that inequality is preserved when any number (positive or negative) is added to both sides.

The statement that $a$ is positive or equal to zero is written $a \geq 0$ (read "a is greater than or equal to zero"). Also, $a \geq b$ means $a > b$ or $a = b$. Hence, $4 \geq 2$ and $4 \geq 4$ are both true inequalities.

When working with inequalities, keep in mind that a product of nonzero numbers is positive if it has an even number of negative factors, or negative if it has an odd number of negative factors.

Solving a linear inequality like

$$3x - 2 < 6 - x$$

means using the above rules to find the values of the variable $x$ for which the inequality is true. For example,

| | |
|---|---|
| $3x - 2 < 6 - x$ | Given. |
| $4x - 2 < 6$ | Add $x$ to both sides. |
| $4x < 8$ | Add 2 to both sides. |
| $x < 2$ | Multiply both side by ¼. |

The solution is the inequality $x < 2$.

The **absolute value** of a number $a$ is denoted by the symbol $|a|$ (read "absolute value of $a$") and is defined by

$$|a| = \begin{cases} a \text{ if } a \geq 0 \\ -a \text{ if } a < 0 \end{cases}$$

For example, $|2| = 2$, $|-6| = -(-6) = 6$, and $|0| = 0$. Clearly, forming an absolute value leaves positive numbers unchanged, and replaces each negative number by the corresponding positive number. The main rules for absolute values are:

- $|a| \geq 0$
- $|ab| = |a||b|$
- $|a + b| \leq |a| + |b|$

On the real line, the absolute value of $a$ is the distance from the point $a$ to the origin, and the distance from $a$ to $b$ is $|a - b|$.

# Problems

1. Solve the linear inequalities $5 - 2x > 17$ and $3x + 4 > 13$.

2. Solve the following equations.

$$|x| = 2 \quad |2x| = 6 \quad |\tfrac{1}{3}x| = 2 \quad |x - 2| = 3 \quad |x + 3| = 1$$

3. By noting that both factors must be positive or both factors must be negative, solve the quadratic inequalities $x(x - 1) > 0$ and $x^2 + 2x - 15 > 0$.

# 7

# Coordinates in a Plane

We can extend the method used to build the real line (page 4) to assign coordinates to points in a plane (Figure 7.1).

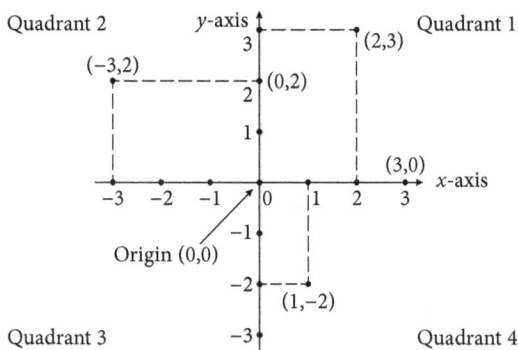

**Figure 7.1** The $xy$-plane

Choose an arbitrary point, called the **origin**, in the plane and draw two perpendicular lines through it, one horizontal and the other vertical. These lines are called the **x-axis** and **y-axis** respectively. Place a coordinate system on each axis so that the unit distance is the same on both. Assign two coordinates to each point $P$ in the plane by drawing perpendicular lines from $P$ to the two axes. These lines intersect the $x$-axis and $y$-axis at points having coordinates $x$ and $y$, and $x$ is called the **x-coordinate** and $y$ is called the **y-coordinate**. Note that if $P$ lies on the $x$-axis, then $y = 0$, and if $P$ lies on the $y$-axis, then $x = 0$. This system establishes a one-to-one correspondence between all points $P$ in the plane and all ordered pairs $(x, y)$ of real numbers.

## The Distance Formula

On the $xy$-plane, points and ordered pairs are essentially identical concepts. This system of assigning coordinates to points in a plane lets us use algebraic tools to study geometry (or more precisely, analytic geometry). Because the same unit of length is used on both axes, we can express the distance between two points in terms of their coordinates.

If $(x_1, y_1)$ and $(x_2, y_2)$ are two points in the $xy$-plane, then the lengths of the legs of the right triangle in Figure 7.2 are $|x_1 - x_2|$ and $|y_1 - y_2|$, where $|a|$ denotes the absolute value of $a$.

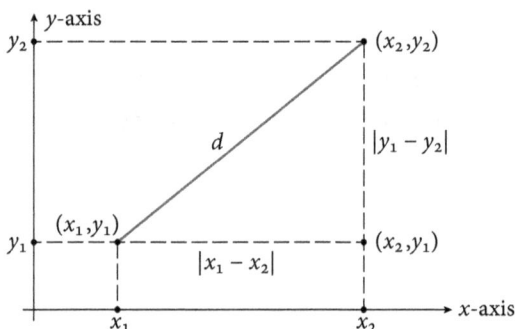

**Figure 7.2** Derivation of the distance formula

Applying the Pythagorean theorem

$$d^2 = |x_1 - x_2|^2 + |y_1 - y_2|^2$$
$$= (x_1 - x_2)^2 + (y_1 - y_2)^2$$

yields the **distance formula**

$$d = \sqrt{(x_1 - x_2)^2 + (y_1 - y_2)^2}$$

In words, the distance between two points equals the square root of the sum of the squares of the differences of their coordinates.

## Problems

1.  In the *xy*-plane, find the length of the sides and the hypotenuse of a right triangle whose vertices are (2, −5), (2, 7), (−3, 7).

2.  In the *xy*-plane, find the area of a rectangle whose vertices are (−3, 7), (4, 2), (−3, 2), (4, 7).

3.  Find the coordinates of the midpoint of the segment joining (1, 2) to (7, 8).

# 8 Functions & Graphs

A **function** associates every number from a first set, called the **domain**, with another number in a second set, called the **range**, such that each element in the domain corresponds to *exactly one* element in the range. (Functions can be defined more generally to deal with other objects, but I consider only real numbers in this book.)

Functions are typically denoted by the letters $f$, $g$, and $h$. If $f$ is a function and $x$ is a number in the domain of $f$, then the number that $f$ associates with $x$ is denoted by $f(x)$ and is called the value of $f$ at $x$. The symbol $f(x)$ is read "$f$ of $x$."

If a function $f$ is defined by the formula

$$f(x) = x^2$$

for every real number $x$, then the domain of $f$ is the set of real numbers, and $f$ is a function that associates every real number with its square. To evaluate $f$ at any number, square that number. For example,

$f(3) = 3^2 = 9$

$f(-\tfrac{1}{2}) = (-\tfrac{1}{2})^2 = \tfrac{1}{4}$

$f(1 + a) = (1 + a)^2 = 1 + 2a + a^2$

Other examples of functions are:

- If a rock is dropped from the edge of a cliff, then it falls $s$ feet in $t$ seconds, and $s$ is a function of $t$. From experiments we know that (approximately) $s = 16t^2$.

- The area $A$ of a circle is a function of its radius $r$. From geometry we know that $A = \pi r^2$.

- If a store owner buys $n$ widgets from a manufacturer at $5 each and the shipping charges are $12, then the owner's total cost $C$ is a function of $n$ given by the formula $C = 5n + 12$.

It's useful to think of a function as a machine or "black box" that when given an input $x$ produces an output $f(x)$ (Figure 8.1). The same input must always produce the same output. Although each input has a unique output, a given output may result from more than one input. The inputs 3 and −3, for example, both produce the output 9 if $f(x) = x^2$.

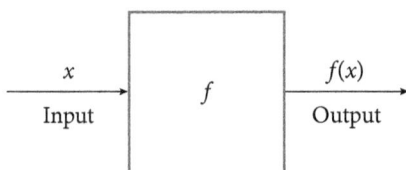

**Figure 8.1** Function as machine

You can also think of a function as a mapping of one set of points (the domain) onto another set of points (the range) (Figure 8.2).

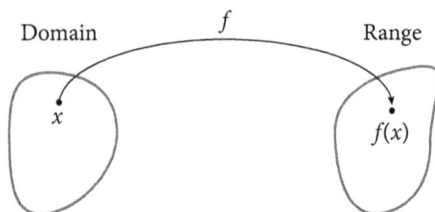

**Figure 8.2** Function as mapping

When working with functions in the $xy$-plane, $y$ is called a function of $x$, symbolized by

$$y = f(x)$$

Using the preceding example, $y = f(x)$ where $f(x) = x^2$. The letter $f$ represents the rule or operation (squaring, in this case) that yields $y$ when applied to $x$.

The best way to visualize a function is by its **graph** (Figure 8.3).

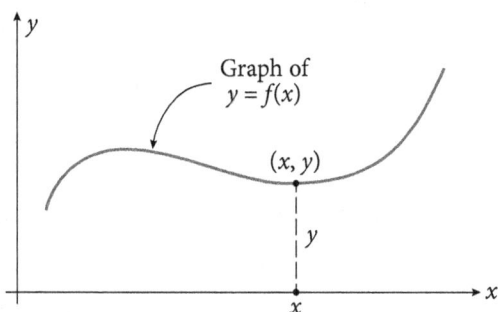

**Figure 8.3** The graph of a function

When a function is defined by an equation in $x$ and $y$, the graph of the function $y = f(x)$ is the set of points $(x, y)$ in the $xy$-plane that satisfies the equation. The variable $x$ is called the **independent variable** because it's free to take on any value in the domain, and $y$ is called the **dependent variable** because its value depends on $x$.

Many types of functions exist, the most familiar being those defined by simple algebraic equations (Figure 8.4). If you're not using a computer, the best way to graph a function is to plot a few "interesting" points and then sketchily connect them according to the characteristics of the equation (linear, power, polynomial, logarithmic, and so on). Graphs for different types of functions are covered in more detail in later chapters.

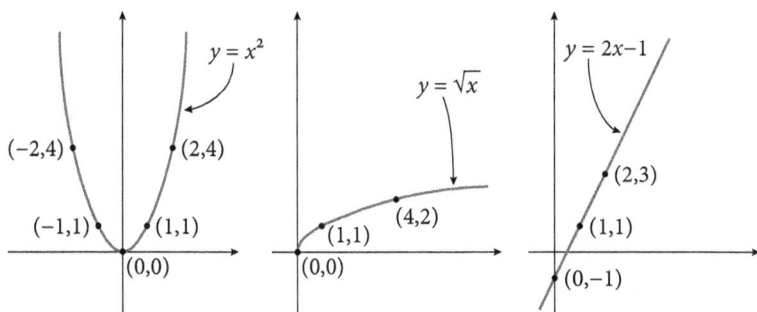

**Figure 8.4** Graphs of a few functions

## Problems

1. If $f(x) = 4x - 3$, find $f(0)$, $f(1)$, $f(2)$, and $f(3)$.

2. If $g(x) = (2x - 4)/(3x^2 + 1)$, find $g(0)$, $g(1)$, and $g(-\frac{1}{2})$.

3. If $h(x) = x^3 - 3x^2 + 5x - 1$, find $h(x^3)$.

4. If $F(x) = x/(x - 1)$, find $F[F(x)]$.

5. Express the area $A$ of a square as a function of the length of one side $s$ and as a function of the perimeter $p$.

6. Express the area $A$ of a circle as a function of its circumference $c$.

7. Express the height $h$ of an equilateral triangle as a function of its base $b$.

8. Two particles start moving along a straight line from the same place at the same time in the same direction. If one particle travels at 40 meters/second and the other at 35 meters/second, find the distance $d$ between them $t$ seconds after they start.

# 9

# Straight Lines

Straight lines are simplest curves in analytic geometry, with the simplest equations.

## The Slope of a Line

If $P_1 = (x_1, y_1)$ and $P_2 = (x_2, y_2)$ are two distinct points on a nonvertical line (Figure 9.1), then the **slope** of the line is denoted by $m$ and defined to be the ratio

$$m = \frac{y_2 - y_1}{x_2 - x_1}$$

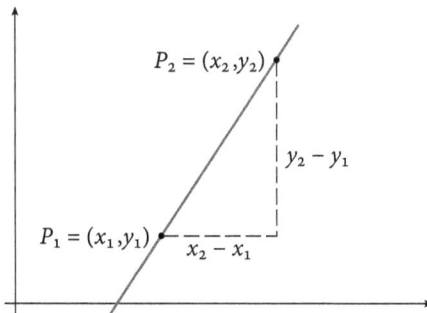

**Figure 9.1** Two points on a straight line

The value of $m$ depends on only the line itself, and not on the particular positions of $P_1$ and $P_2$. In the figure above, $m$ is the ratio of the height of the indicated triangle to its base.

If $P_2$ is assumed to be a point on the line one unit to the right of $P_1$, so that $x_2 - x_1 = 1$, then

$$m = \frac{y_2 - y_1}{1} = y_2 - y_1$$

This result shows that if we move one unit directly to the right from any point on the line, the slope $m$ can be thought of as the distance up or down that we must move to get back on the line. Thus, the sign of the slope is related as follows to the direction of the line:

- If $m > 0$, then the line rises to the right.
- If $m < 0$, then the line falls to the right.
- If $m = 0$, then the line is horizontal.

Also, the numerical magnitude (absolute value) of the slope is a measure of the steepness of the line (Figure 9.2).

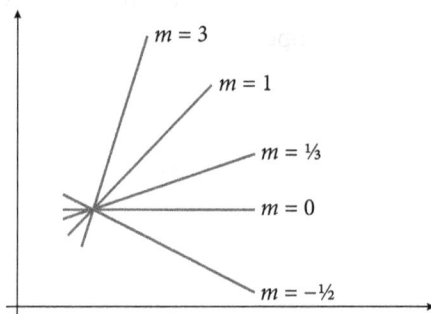

**Figure 9.2** Examples of various slopes

## Equations of Straight Lines

The three main forms for the equation of a straight line are

$$x = a \qquad\qquad \frac{y - y_0}{x - x_0} = m \qquad\qquad y = mx + b$$

- The first equation specifies a **vertical line**, where all points have the same $x$-coordinate, and the $y$-coordinate is irrelevant (Figure 9.3).
- The second equation, or **point-slope equation**, specifies a nonvertical line by a point $(x_0, y_0)$ on the line, and its slope $m$ (Figure 9.4).

This equation states an algebraic condition that's satisfied by the indicated variable point $(x, y)$ on the line.

- The third equation, or **slope-intercept equation**, is a special case of the second that we get when the given point is a point $(0, b)$ on the $y$-axis (Figure 9.5). The number $b$ is called the **$y$-intercept** of the line.

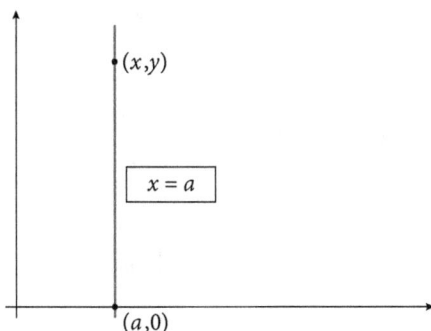

**Figure 9.3** Equation for a vertical line

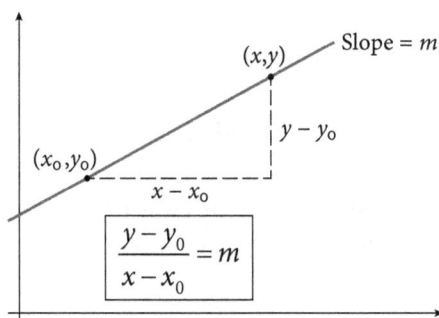

**Figure 9.4** The point-slope equation

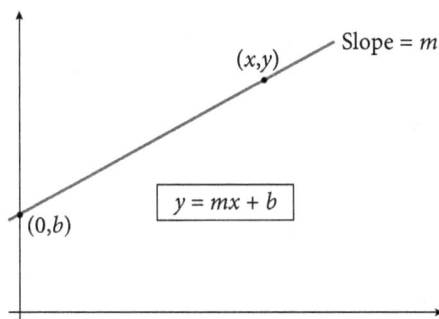

**Figure 9.5** The slope-intercept equation

## Parallel and Perpendicular Lines

Two lines with slopes $m_1$ and $m_2$ are parallel if and only if their slopes are equal: $m_1 = m_2$. If two lines are perpendicular, then

$$m_1 m_2 = -1$$

Which isn't particularly obvious, but can be understood with the aid of similar triangles (Figure 9.6).

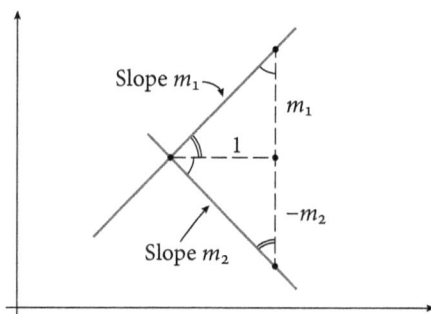

**Figure 9.6** Perpendicular lines

Assume that the lines in the figure above are perpendicular. Draw a line segment of length 1 rightward from their point of intersection, and from its right endpoint draw vertical line segments up and down to the two perpendicular lines. From the definition of the slopes, we know that the two right triangles formed this way have sides of the indicated lengths. Because the lines are perpendicular, the indicated angles are equal, and therefore the triangles are similar and the following ratios of their corresponding sides are equal:

$$\frac{m_1}{1} = \frac{1}{-m_2}$$

which is equivalent to $m_1 m_2 = -1$. (By reversing this reasoning, it's clear that this condition also implies perpendicularity.)

## Problems

1. Sketch diagrams of $x = 3$ and $y = -4$. Sketch a diagram of all points that satisfy $x < 3$ and $y > 2$.

2. Sketch diagrams of $x$ or $y$ (or both) is zero; of $x \geq 0$ and $y \leq 0$; and of $x^2 \leq 1$.

3. Find the slopes of the lines determined by the following pairs of points.

   (1, 2) and (3, 6)

   (-2, 4) and (3, 4)

   (-2, 1) and (2, -3)

4. Show that the lines $3x + y = 2$ and $2y = 1 - 6x$ are parallel.

5. Write the equation of the line through (-2, -3) which is parallel to $x + 2y = 3$ and which is perpendicular to $x + 2y = 3$.

# 10

The **circle** with center $C$ and radius $r$ is defined to be the set of all points whose distance from $C$ is $r$. If the coordinates of $C$ are $(h, k)$ and the coordinates of a variable point on the circle are $(x, y)$, then by the distance formula (page 30), the condition is

$$\sqrt{(x-h)^2 + (y-k)^2} = r$$

Squaring to remove the radical yields the standard equation of the circle:

$$(x - h)^2 + (y - k)^2 = r^2$$

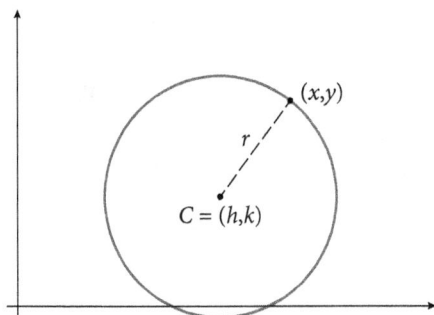

**Figure 10.1** A circle in the $xy$-plane

In particular,

$$x^2 + y^2 = r^2$$

is the equation of the circle with radius $r$ and center at the origin, and the equation $x^2 + y^2 = 1$ is the **unit circle**.

The circle with center $(-3, 2)$ and radius 5, for example, has the equation $(x + 3)^2 + (y - 2)^2 = 25$.

The equation $(x - 4)^2 + (y + 2)^2 = 12$ is easily recognized as the equation of the circle with center $(4, -2)$ and radius $\sqrt{12} = 2\sqrt{3}$.

Note that expanding $(x - 4)^2 + (y + 2)^2 = 12$ by squaring and collecting terms actually makes it harder to recognize:

$$x^2 + y^2 + 6x - 4y - 12 = 0$$

To return this expanded form to standard form, you would have to complete the square (Chapter 5) on the variables $x$ and $y$.

To change the equation $x^2 + y^2 - 10x + 6y + 18 = 0$, for example, to standard form, prepare it for completing the square:

$$(x^2 - 10x + ?) + (y^2 + 6y + ?) = -18$$

The numbers that must be added inside the parentheses to make these quantities perfect squares are 25 and 9, so the equation becomes

$$(x^2 - 10x + 25) + (y^2 + 6y + 9) = -18 + 25 + 9$$

or equivalently,

$$(x - 5)^2 + (y + 3)^2 = 16$$

which is the equation of the circle with center $(5, -3)$ and radius 4.

Note that if the constant term 18 in the first equation in this example is replaced by 34, then the final equation has 0 on the right side and is therefore the equation of a single point $(5, -3)$, which can be thought of as a circle with center $(5, -3)$ and radius $r = 0$.

# Problems

1. Write the equation of the circle with center $(0, 0)$ and radius 2; with center $(-2, 0)$ and radius 7; with center $(3, 6)$ and radius ½; and with $(5, 5)$ and $(-3, -1)$ as the ends of a diameter.

2. In the $xy$-plane, sketch both the line $x + 16 = 7y$ and the circle $x^2 + y^2 - 4x + 2y = 20$ and find their points of intersection.

3. Find the center and radius of the following circles.

$$(x+3)^2 + (y-6)^2 = 9$$

$$(x-4)^2 + y^2 = 4$$

$$x^2 + (y+2)^2 = 1$$

$$x^2 + y^2 + 6x + 2y + 6 = 0$$

$$x^2 + y^2 - 16x + 14y + 97 = 0$$

# 11

# Parabolas

A **parabola** is a plane curve that's the graph of a quadratic function. Each point on a parabola is equidistant from a given fixed point, called the **focus**, and a given fixed line, called the **directrix**. To find the simplest equation for this curve, let the focus be the point $F = (0, p)$ where $p$ is a positive number, and let the directrix be the line $y = -p$ (Figure 11.1).

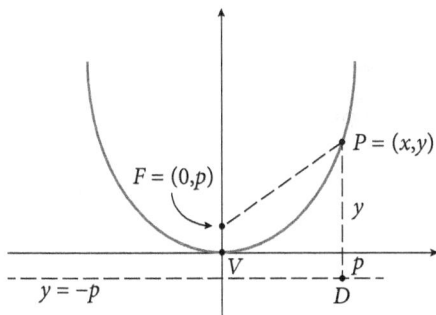

**Figure 11.1** A parabola, opening up

If $P = (x, y)$ is any point on the parabola, then $PF = PD$ by definition. By applying the distance formula (page 30), this condition becomes

$$\sqrt{x^2 + (y - p)^2} = y + p$$

Squaring and simplifying yields

$$x^2 = 4py \quad \text{or} \quad y = \frac{1}{4p}x^2$$

as the equation of this parabola.

The line through the focus perpendicular to the directrix is called the **axis** of the parabola, and the point $V$ where the parabola intersects the axis is called the **vertex**.

The parabola in Figure 11.1 opens up. If the focus is $F = (0, -p)$ and directrix is $y = p$, however, with $p$ still a positive number, then the parabola opens down (Figure 11.2) and a similar calculation shows that its equation is now

$$x^2 = -4py \quad \text{or} \quad y = -\frac{1}{4p}x^2$$

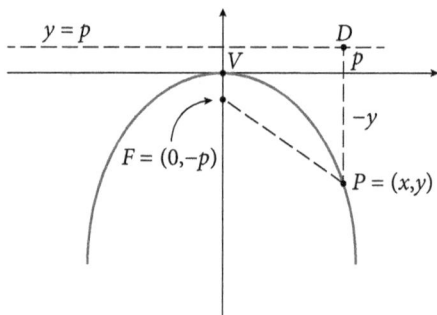

**Figure 11.2** A parabola, opening down

For both parabolas (up and down) the positive number $p$ is the distance from the vertex to the focus. In each case the parabola is symmetric about its axis, as can be seen geometrically and also from the fact that each equation is unchanged when $x$ is replaced by $-x$. Also note that the vertex is the low point of the parabola when it opens up and the high point when it opens down.

The equation $y = x^2$ is of the first type with $4p = 1$, so $p = \frac{1}{4}$. It therefore represents the parabola with vertex at the origin which opens up and has focus $(0, \frac{1}{4})$ and directrix $y = -\frac{1}{4}$.

To see another point about parabolas, consider the equation

$$y = x^2 - 6x + 11$$

Write this equation in the form

$$y - 11 = x^2 - 6x$$

and then complete the square (Chapter 5) on the terms involving $x$ to get

$$y - 2 = (x - 3)^2$$

Now introduce the new variables $x_0$ and $y_0$ as

$$x_0 = x - 3$$

$$y_0 = y - 2$$

The original equation now becomes

$$y_0 = (x_0)^2$$

The graph of this equation is a parabola with vertex at the origin of the $x_0$, $y_0$ coordinate system, and this origin is located at the point (3, 2) in the $x$, $y$ system (Figure 11.3).

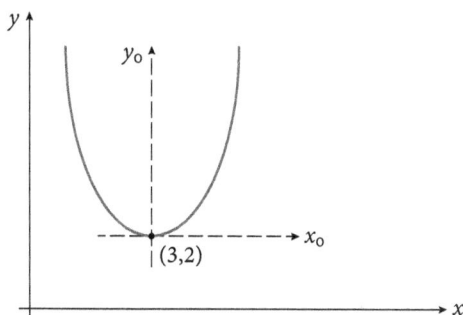

**Figure 11.3** Parabola shifted from the $x$, $y$ origin

In the same way, any equation of the form

$$y = ax^2 + bx + c \qquad (a \neq 0)$$

represents a parabola with vertical axis that opens up if $a > 0$ or down if $a < 0$. The vertex of this parabola is located by completing the square on $x$, and the equation can be written in the form

$$(x - h)^2 = 4p(y - k)$$

where the point $(h, k)$ is the vertex.

## Problems

1. Find the focus and directrix of each of the following parabolas.

   $$y = 2x^2 \qquad y = \frac{1}{8}x^2$$

   $$y = -5x^2 \qquad y = -\frac{1}{12}x^2$$

2. A parabola has vertical axis and vertex at the origin. Write its equation if its focus is at (0, 3); (0, 16); (0, –1); and (0, –1/10).

3. Find the vertex and focus of each of the following parabolas, and indicate whether it opens up or down.

   $$y = x^2 - 4x + 1 \qquad y = 2x^2 - 12x - 7$$

   $$y = -x^2 - 4x + 5 \qquad y = 4 - 2x - \frac{1}{2}x^2$$

# 12      Types of Functions

This chapter covers a few of the simpler functions and the characteristic features of their graphs.

## Linear Functions

**Linear functions** are of the form

$$y = f(x) = ax + b$$

which is the equation of a straight line with slope $a$ and $y$-intercept $b$ (Figure 12.1, left). If $a = 0$, we have a **constant function** (Figure 12.1, right).

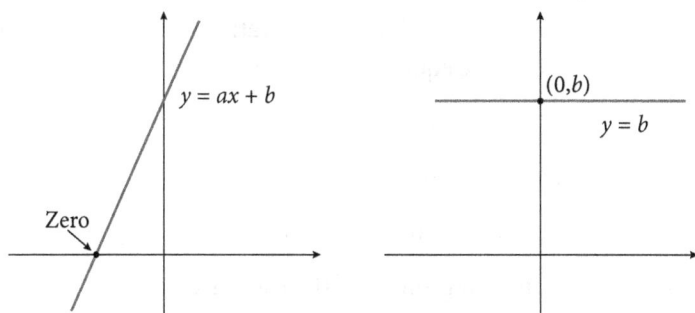

**Figure 12.1** Linear functions

For any function $y = f(x)$, a value of $x$ for which $f(x) = 0$ is called a **zero** of the function. The zeros of a linear function are therefore the $x$-intercepts of its graph, or equivalently, the roots of the equation $f(x) = 0$. Solving a linear equation by finding the zero of a nonconstant linear function is covered in Chapter 5.

## Quadratic Functions

**Quadratic functions** are of the form

$$y = f(x) = ax^2 + bx + c \qquad (a \neq 0)$$

Two real roots | One real root | Two imaginary roots

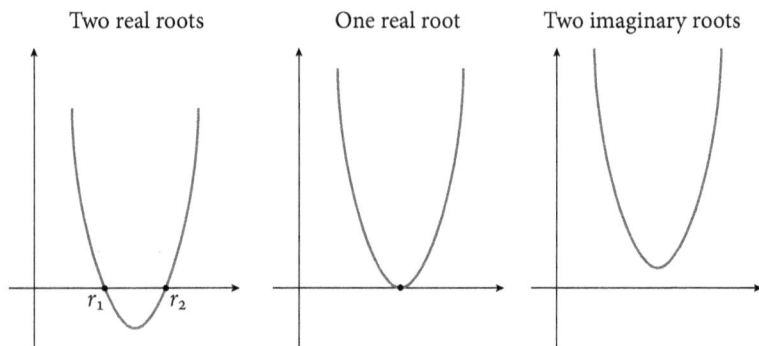

**Figure 12.2** Quadratic functions

If $a > 0$, then this function's graph is a parabola that opens up (Chapter 11). If the graph is low enough, then the function has two zeros $r_1$ and $r_2$ (Figure 12.2, left). These are the roots of the quadratic equation

$$ax^2 + bx + c = 0 \quad \text{or} \quad x^2 + \frac{b}{a}x + \frac{c}{a} = 0$$

and can always be found by using the quadratic formula. In terms of its roots, the second of these equations can be written in factored form as

$$(x - r_1)(x - r_2) = 0$$

which multiplied out becomes

$$x^2 - (r_1 + r_2)x + r_1 r_2 = 0$$

Solving, the sum and product of the roots are

$$r_1 + r_2 = -b/a \quad \text{and} \quad r_1 r_2 = c/a$$

The cases of a single real root and two imaginary roots (Figure 12.2) are mentioned in Chapter 5. In terms of the quadratic formula

$$x = \frac{-b \pm \sqrt{b^2 - 4ac}}{2a}$$

these cases correspond to the conditions $b^2 - 4ac = 0$ and $b^2 - 4ac < 0$.

Polynomial functions of higher degree can be hard to sketch. Figure 12.3, however, shows the graph of the **power function**

$$y = f(x) = x^n$$

when $n$ is even and $\geq 2$, and when $n$ is odd and $\geq 3$. For larger values of $n$ these curves are flatter near $x = 0$ and steeper for $|x| > 1$.

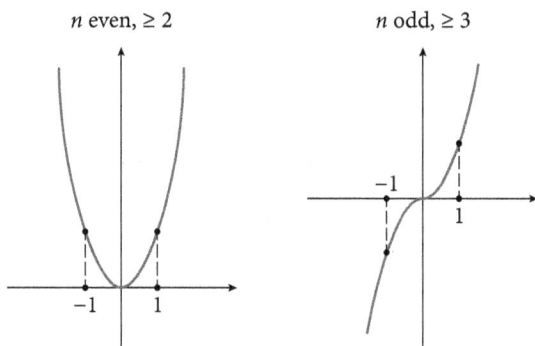

**Figure 12.3** Power functions

## Radical Functions

Figure 12.4 shows the graph of the square root function

$$y = \sqrt{x}$$

This curve is the top half of the graph of the equation $y^2 = x$, which is a parabola opening to the right.

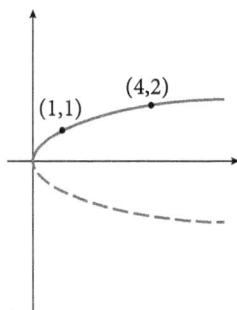

**Figure 12.4** Square root function

Figure 12.5 shows the graph of

$$y = \sqrt{4 - x^2}$$

which is the top half of the circle $x^2 + y^2 = 4$.

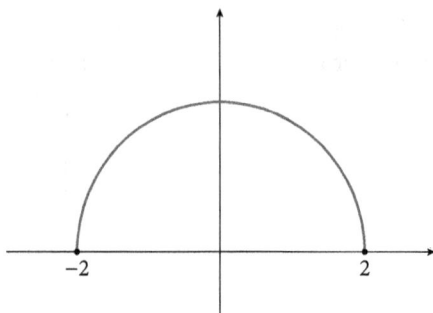

**Figure 12.5** Semicircle function

## Rational Functions

Just as rational numbers are quotients of integers, **rational functions** are quotients of polynomials. Figure 12.6 shows the graphs of the rational functions

$$y = \frac{1}{x} \qquad y = \frac{1}{x^2 - 2x} = \frac{1}{x(x-2)} \qquad y = \frac{1}{x^2 + 1}$$

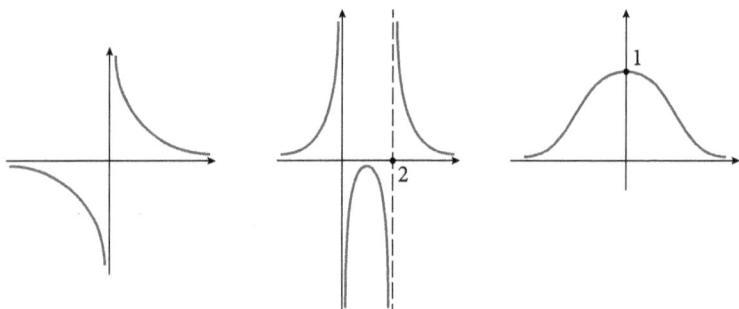

**Figure 12.6** Rational functions

Examine these graphs, paying special attention to the zeros of the denominator of a rational function. The best method for sketching a graph is to learn the characteristic features of various functions (zeros, turning points, and so on) and use these features as the basis for the sketch. (As a last resort, you can plot a few points and connect them.)

## Problems

1. Without actually finding the roots, determine for each of the following equations whether its roots are real and distinct, real and equal, or imaginary.

$$5x^2 + 3x + 4 = 0 \qquad 7x^2 - 2x - 15 = 0$$

$$3x^2 - 5x + 1 = 0 \qquad 4x^2 + 20x + 25 = 0$$

2. Without solving, find the sum and product of the roots of the following equations.

$$4x^2 - 7x - 13 = 0$$

$$3x^2 + 10x + 17 = 0$$

$$2x^2 + x - 2 = 0$$

3. Construct a quadratic equation having the following numbers as roots.

$$-3, 8$$

$$2 + \sqrt{5}, 2 - \sqrt{5}$$

$$-\frac{3}{5}, \frac{2}{3}$$

4. Sketch a graph of the following equations.

$$f(x) = x^2 - 4x + 4 \qquad f(x) = x^2 - x - 2$$

$$f(x) = \frac{1}{x^2} \qquad\qquad f(x) = \frac{x}{x^2 + 1}$$

# 13            Logarithms

Logarithms are exponents (Chapter 2). In the equation $100 = 10^2$, for example, the exponent 2 is the logarithm of 100 to the base 10. To understand logarithms, first note that the equations

$$x = 2y \quad \text{and} \quad y = \tfrac{1}{2}x$$

are in fact only one equation that expresses the same relation between $x$ and $y$, written first in a form solved for $x$ and second in a form solved for $y$. Similarly, if $a$ is any constant $> 1$, then the equations

$$x = a^y \quad \text{and} \quad y = \log_a x$$

are equivalent, except that the second equation is solved for $y$ and the symbol "$\log_a$" denotes this operation. The following examples show equivalent statements about exponents (on the left) and logarithms (on the right).

$$1000 = 10^3 \qquad\qquad 3 = \log_{10} 100$$

$$16 = 2^4 \qquad\qquad 4 = \log_2 16$$

$$2 = 8^{1/3} \qquad\qquad \frac{1}{3} = \log_8 2$$

$$\frac{1}{9} = 3^{-2} \qquad\qquad -2 = \log_3 \frac{1}{9}$$

$$1 = a^0 \qquad\qquad 0 = \log_a 1$$

The rules for logarithms are derived from the corresponding rules for exponents:

| Rule | Proof |
|------|-------|

$$\log_a x_1 x_2 = \log_a x_1 + \log_a x_2 \qquad x_1 x_2 = a^{y_1} \cdot a^{y_2} = a^{y_1 + y_2}$$

$$\log_a \frac{x_1}{x_2} = \log_a x_1 - \log_a x_2 \qquad \frac{x_1}{x_2} = \frac{a^{y_1}}{a^{y_2}} = a^{y_1 - y_2}$$

$$\log_a x^n = n \log_a x \qquad x^n = (a^y)^n = a^{ny}$$

To prove the first rule in detail, note that if

$$x_1 = a^{y_1} \quad \text{and} \quad x_2 = a^{y_2}$$

or equivalently,

$$y_1 = \log_a x_1 \quad \text{and} \quad y_2 = \log_a x_2$$

then by the rule of exponents,

$$x_1 x_2 = a^{y_1} \cdot a^{y_2} = a^{y_1 + y_2}$$

The exponent on the right, which is $\log_a x_1 x_2$, is also $\log_a x_1 + \log_a x_2$, and this is the first rule. The other two rules are proved similarly.

Two more useful facts about logarithms are

$$a^{\log_a x} = x \quad \text{and} \quad \log_a a^x = x$$

The first fact means that when $x$ is expressed as some power of $a$, the exponent *is* $\log_a x$. The second fact means that the left side *is* the exponent when $a^x$ is expressed as some power of $a$.

If you need to shift logarithms from one base to another, note that

$$x = b^y = \left( a^{\log_a b} \right)^y = a^{y \log_a b}$$

which translates to

$$\log_a x = (\log_a b)(\log_b x)$$

In most applications, logarithms to the base $e$ or base 2 are used. Base 10 rarely appears in numerical calculations.

# Problems

1. Express each of the following equations in terms of logarithms.

$$5^2 = 25 \quad 2^5 = 32 \quad 5^{-2} = \frac{1}{25} \quad 81^{0.5} = 9$$

$$7^0 = 1 \quad 10^{-1} = \frac{1}{10} \quad 32^{4/5} = 16 \quad 16^{0.75} = 8$$

2. Express each of the following equations in terms of exponents.

$$\log_{10} 10 = 1 \quad \log_4 8 = \frac{3}{2} \quad \log_2 8 = 3 \quad \log_5 \frac{1}{125} = -3$$

$$\log_{10} 0.01 = -2 \quad \log_7 343 = 3 \quad \log_5 125 = 3 \quad \log_{10} 0.1 = -1$$

3. Evaluate the following logarithmic expressions.

$$\log_2 2 \quad \log_{10} 10000 \quad \log_2 16 \quad\quad \log_{25} 125$$

$$\log_8 4 \quad \log_2 1 \quad\quad \log_8 \frac{\sqrt{2} \cdot \sqrt[3]{256}}{\sqrt[6]{32}} \quad \log_9 \frac{81(9)^{4/3}}{\sqrt[3]{9^2} \cdot \sqrt{9^3}}$$

4. Solve for $x$.

$$\log_9 x = 3.5 \quad \log_{27} x = \frac{5}{3} \quad \log_2 x = 8 \quad \log_{32} x = 0.8$$

5. Find the base $a$.

$$\log_a 9 = 0.4 \quad \log_a 27 = -\frac{3}{4} \quad \log_a 49 = 2 \quad \log_a 6 = \frac{1}{2}$$

# 14   Dividing Polynomials

In arithmetic, division is often applied to reduce an improper fraction to the sum of an integer and a proper fraction, as in

$$\frac{17}{3} = 5 + \frac{2}{3}$$

In algebra, division of polynomials is applied in a similar way to a rational function (quotient of polynomials) if the degree of the numerator is greater than or equal to the degree of the denominator. For example,

$$\frac{2x^3 + 3x^2 + 7x - 1}{x^2 + 1} = 2x + 3 + \frac{5x - 4}{x^2 + 1}$$

In effect, the "improper" rational function on the left is reduced to a polynomial plus a "proper" rational function with the same denominator as the original rational function.

In general terms, if we're given a rational function

$$\frac{f(x)}{g(x)}$$

in which $\deg f(x) \geq \deg g(x)$ ["deg" means degree], then dividing $f(x)$ by $g(x)$ means finding polynomials $Q(x)$ and $R(x)$ such that

$$\frac{f(x)}{g(x)} = Q(x) + \frac{R(x)}{g(x)}$$

where $\deg R(x) < \deg Q(x)$. By convention, $f(x)$ is the **dividend**, $g(x)$ is the **divisor**, $Q(x)$ is the **quotient**, and $R(x)$ is the **remainder**.

To understand how polynomial division is done, first consider how it's carried out in the arithmetic example above, which is equivalent to

$$17 = 5 \cdot 3 + 2 \quad \text{or} \quad 17 - 5 \cdot 3 = 2$$

The second of these equations shows the procedure: subtract from 17 (the dividend) a multiple of 3 (the divisor) that leaves a remainder of 2, which is as small as possible without being negative. In the same way, the algebraic example above is equivalent to

$$(2x^3 + 3x^2 + 7x - 1) - (2x + 3)(x^2 + 1) = 5x - 4$$

This equation shows the process for dividing polynomials, which is basically the process of finding the quotient $2x + 3$. Note that the first term $2x^3$ of the dividend is the product of $2x$, the first term of the quotient, and $x^2$, the first term of the divisor. The steps are:

1. Find the first term of the quotient by dividing the first term of the dividend by the first term of the divisor, where in each polynomial the terms are written in order of decreasing exponents.

2. Multiply the divisor by the first term of the quotient that you just found.

3. Subtract the result from the dividend, which reduces the degree of the remaining polynomial.

4. Repeat the preceding steps until the residual polynomial has degree less than that of the divisor.

This method is called **long division**:

$$
\begin{array}{r}
2x + 3 \qquad \qquad \text{QUOTIENT} \\
\text{DIVISOR} \quad x^2 + 1 \overline{)\, 2x^3 + 3x^2 + 7x - 1} \quad \text{DIVIDEND} \\
\underline{2x^3 \qquad\quad + 2x} \\
3x^2 + 5x - 1 \\
\underline{3x^2 \qquad\quad + 3} \\
5x - 4 \quad \text{REMAINDER}
\end{array}
$$

The division equation given above can be written in the form

$$f(x) = Q(x)g(x) + R(x) \quad \text{or} \quad f(x) - Q(x)g(x) = R(x)$$

and in any specific case can be carried out in just the same way.

To divide $2x^3 + x^2 - 13x + 10$ by $x - 2$, for example,

$$
\begin{array}{r}
2x^2 + 5x\ -3 \\
x - 2 \overline{)\,2x^3 +\ x^2 - 13x + 10} \\
\underline{2x^3 - 4x^2} \\
5x^2 - 13x + 10 \\
\underline{5x^2 - 10x} \\
-\ 3x + 10 \\
\underline{-\ 3x +\ 6} \\
4
\end{array}
$$

This result can be written as

$$
\frac{2x^3 + x^2 - 13x + 10}{x - 2} = 2x^2 + 5x - 3 + \frac{4}{x - 2}
$$

or as

$$
2x^3 + x^2 - 13x + 10 = (2x^2 + 5x - 3)(x - 2) + 4
$$

If a polynomial $f(x)$ is divided by a linear polynomial $x - r$, then the quotient is a polynomial $Q(x)$ and the remainder is a constant $R$:

$$
f(x) = (x - r)Q(x) + R
$$

To solve for $R$, set $x = r$, yielding $f(r) = 0 + R$ or $R = f(r)$. This result says that the remainder when $f(x)$ is divided by $x - r$ is the value of $f(x)$ at $x = r$. This result is summarized as the:

**Remainder Theorem.** If $f(x)$ is a polynomial and $r$ is any real number, then the remainder when $f(x)$ is divided by $x - r$ is $f(r)$; that is,

$$
f(x) = (x - r)Q(x) + f(r)
$$

where $Q(x)$ is a polynomial.

Note from this theorem that $f(r) = 0$ if and only if $f(x) = (x - r)Q(x)$, that is, if and only if $x - r$ is a factor of $f(x)$. This result is the:

**Factor Theorem.** If $f(x)$ is a polynomial, then a real number $r$ is a root of the equation $f(x) = 0$ if and only if $x - r$ is a factor of $f(x)$.

## Problems

1. Use long division to find the quotient $Q(x)$ and the remainder $R(x)$.

$$\frac{x^2 - 5x + 6}{x - 3}$$

$$\frac{x^3 - 8x^2 + x + 42}{x - 7}$$

$$\frac{3x^4 - 2x^3 + 10x^2 - 7x + 10}{x^2 + 3}$$

$$\frac{x^5 - 4x^4 + 3x^3 + 7x^2 - 10x - 5}{x^2 - 2x - 1}$$

2. Factor completely the following polynomials $f(x)$, using the given root $x$.

$$x^3 - x^2 + x - 1, x = 1$$

$$x^3 + 5x^2 - 12x - 36, x = -2$$

$$x^3 - x^2 - 5x - 3, x = 3$$

3. If $n$ is any positive integer, use the factor theorem to show that $x^n - 1$ has $x - 1$ as a factor, and then use long division to compute the other factor.

# 15

# Systems of Linear Equations

This chapter provides a taste of some topics in systems of equations. Proper treatment of this subject requires a full course in linear algebra.

The symbol

$$\begin{vmatrix} a_1 & b_1 \\ a_2 & b_2 \end{vmatrix}$$

is called a **second-order determinant**. It's used to denote the number $a_1b_2 - b_1a_2$, so by definition

$$\begin{vmatrix} a_1 & b_1 \\ a_2 & b_2 \end{vmatrix} = a_1b_2 - b_1a_2$$

The number $a_1b_2 - b_1a_2$ is called the **value** or **expansion** of the determinant, and the numbers $a_1$, $b_1$, $a_2$, and $b_2$ are called its **elements**. For example,

$$\begin{vmatrix} 2 & -6 \\ -3 & 7 \end{vmatrix} = 2 \cdot 7 - (-6) \cdot (-3) = 14 - 18 = -4$$

You can remember the value of a second order determinant as the difference of the products of the diagonal elements, illustrated as

$$+\begin{vmatrix} a_1 & b_1 \\ a_2 & b_2 \end{vmatrix}-$$

Second-order determinants can be used to define convenient formulas for the solution of a system of two linear equations in two unknowns. To solve the system (shown here in standard form)

$$a_1x + b_1y = c_1$$
$$a_2x + b_2y = c_2$$

first eliminate $y$ by multiplying the equations by $b_2$ and $b_1$, respectively, and subtracting. Next, eliminate $x$ by multiplying the equations by $a_2$ and $a_1$, respectively, and subtracting. The result of these procedures is

$$(a_1b_2 - b_1a_2)x = c_1b_2 - c_2b_1$$
$$(a_1b_2 - b_1a_2)y = a_1c_2 - a_2c_1$$

Solve for $x$ and $y$ to get

$$x = \frac{c_1b_2 - b_1c_2}{a_1b_2 - b_1a_2} \qquad y = \frac{a_1c_2 - c_1a_2}{a_1b_2 - b_1a_2}$$

where $a_1b_2 - b_1a_2 \neq 0$. In determinant form, the values of $x$ and $y$ are

$$x = \frac{\begin{vmatrix} c_1 & b_1 \\ c_2 & b_2 \end{vmatrix}}{\begin{vmatrix} a_1 & b_1 \\ a_2 & b_2 \end{vmatrix}} \qquad y = \frac{\begin{vmatrix} a_1 & c_1 \\ a_2 & c_2 \end{vmatrix}}{\begin{vmatrix} a_1 & b_1 \\ a_2 & b_2 \end{vmatrix}}$$

In each case ($x$ and $y$), the denominator is the determinant of the array of coefficients on the left side of the original system above. In the formula for $x$, the numerator is the determinant in the denominator with the column of $x$-coefficients replaced by the column of constants on the right side of the original system. In the formula for $y$, the numerator has the column of $y$-coefficients replaced by the column of constants.

For example, to solve the system

$$4y = x - 2$$
$$3x = 4 - y$$

first rearrange the equations into standard form

$$x - 4y = 2$$
$$3x + y = 4$$

and then solve for $x$ and $y$ by using determinants. The solution to the system is

$$x = \frac{\begin{vmatrix} 2 & -4 \\ 4 & 1 \end{vmatrix}}{\begin{vmatrix} 1 & -4 \\ 3 & 1 \end{vmatrix}} = \frac{2 \cdot 1 - (-4) \cdot 4}{1 \cdot 1 - (-4) \cdot 3} = \frac{2+16}{1+12} = \frac{18}{13}$$

$$y = \frac{\begin{vmatrix} 1 & 2 \\ 3 & 4 \end{vmatrix}}{\begin{vmatrix} 1 & -4 \\ 3 & 1 \end{vmatrix}} = \frac{1 \cdot 4 - 2 \cdot 3}{1 \cdot 1 - (-4) \cdot 3} = \frac{-2}{1+12} = -\frac{2}{13}$$

Geometrically, the simultaneous solution of this system gives the coordinates of the point in the $xy$-plane at which the straight lines $4y = x - 2$ and $3x = 4 - y$ intersect.

**Third-order determinants** are used to solve systems of three linear equations in three unknowns. Here, the value of a third-order determinant is defined by means of the elements in its first row and certain corresponding second-order determinants, as follows:

$$\begin{vmatrix} a_1 & b_1 & c_1 \\ a_2 & b_2 & c_2 \\ a_3 & b_3 & c_3 \end{vmatrix} = a_1 \begin{vmatrix} b_2 & c_2 \\ b_3 & c_3 \end{vmatrix} - b_1 \begin{vmatrix} a_2 & c_2 \\ a_3 & c_3 \end{vmatrix} + c_1 \begin{vmatrix} a_2 & b_2 \\ a_3 & b_3 \end{vmatrix}$$

This logic behind this mysterious definition can be made clear by an in-depth linear algebra class, but a few elementary facts explain some of the mathematical reasoning.

The second-order determinants in this equation are called **minors** of the elements $a_1$, $b_1$, and $c_1$. In general, the minor of any element of the third-order determinant on the left side of the definition is the second-order determinant that remains after deleting the row and column containing the element of interest.

Thus the minors of $a_1$ and $b_1$ are

$$\begin{vmatrix} a_1 & b_1 & c_1 \\ a_2 & b_2 & c_2 \\ a_3 & b_3 & c_3 \end{vmatrix} = \begin{vmatrix} b_2 & c_2 \\ b_3 & c_3 \end{vmatrix} \quad \text{and} \quad \begin{vmatrix} a_1 & b_1 & c_1 \\ a_2 & b_2 & c_2 \\ a_3 & b_3 & c_3 \end{vmatrix} = \begin{vmatrix} a_2 & c_2 \\ a_3 & c_3 \end{vmatrix}$$

Another fact that must remain a mystery until in-depth study: the value of the third-order determinant as defined above can also be found by multiplying the elements of *any* row (or column) by their respective minors, with sign attached to the terms according to this pattern:

$$\begin{vmatrix} + & - & + \\ - & + & - \\ + & - & + \end{vmatrix}$$

With this fact in hand, note that the definition of the value of a third-order determinant as given above is just one among many possible expansions. The expansion by the second column, for example, is

$$\begin{vmatrix} a_1 & b_1 & c_1 \\ a_2 & b_2 & c_2 \\ a_3 & b_3 & c_3 \end{vmatrix} = -b_1 \begin{vmatrix} a_2 & c_2 \\ a_3 & c_3 \end{vmatrix} + b_2 \begin{vmatrix} a_1 & c_1 \\ a_3 & c_3 \end{vmatrix} - b_3 \begin{vmatrix} a_1 & c_1 \\ a_2 & c_2 \end{vmatrix}$$

The theory of determinants (omitted here) guarantees that each method of expansion described above yields the same result. The value of the following determinant, for example, calculated by using the first row is

$$\begin{vmatrix} 1 & -2 & 4 \\ 3 & -1 & 6 \\ 2 & 3 & 2 \end{vmatrix} = 1 \cdot \begin{vmatrix} -1 & 6 \\ 3 & 2 \end{vmatrix} -(-2) \cdot \begin{vmatrix} 3 & 6 \\ 2 & 2 \end{vmatrix} + 4 \cdot \begin{vmatrix} 3 & -1 \\ 2 & 3 \end{vmatrix}$$

$$= (-2 - 18) + 2(6 - 12) + 4(9 + 2) = 12$$

and then by using the second column is

$$\begin{vmatrix} 1 & -2 & 4 \\ 3 & -1 & 6 \\ 2 & 3 & 2 \end{vmatrix} = -(-2)\cdot\begin{vmatrix} 3 & 6 \\ 2 & 2 \end{vmatrix} + (-1)\cdot\begin{vmatrix} 1 & 4 \\ 2 & 2 \end{vmatrix} - 3\cdot\begin{vmatrix} 1 & 4 \\ 3 & 6 \end{vmatrix}$$

$$= 2(6-12)-(2-8)-3(6-12) = 12$$

To solve a system of three linear equations in three unknowns (shown here in standard form)

$$a_1x + b_1y + c_1z = d_1$$
$$a_2x + b_2y + c_2z = d_2$$
$$a_3x + b_3y + c_3z = d_3$$

use the standard algebraic methods of elimination and express the solution by means of determinants in the form

$$x = \frac{\begin{vmatrix} d_1 & b_1 & c_1 \\ d_2 & b_2 & c_2 \\ d_3 & b_3 & c_3 \end{vmatrix}}{\begin{vmatrix} a_1 & b_1 & c_1 \\ a_2 & b_2 & c_2 \\ a_3 & b_3 & c_3 \end{vmatrix}} \qquad y = \frac{\begin{vmatrix} a_1 & d_1 & c_1 \\ a_2 & d_2 & c_2 \\ a_3 & d_3 & c_3 \end{vmatrix}}{\begin{vmatrix} a_1 & b_1 & c_1 \\ a_2 & b_2 & c_2 \\ a_3 & b_3 & c_3 \end{vmatrix}} \qquad z = \frac{\begin{vmatrix} a_1 & b_1 & d_1 \\ a_2 & b_2 & d_2 \\ a_3 & b_3 & d_3 \end{vmatrix}}{\begin{vmatrix} a_1 & b_1 & c_1 \\ a_2 & b_2 & c_2 \\ a_3 & b_3 & c_3 \end{vmatrix}}$$

where the determinant in the denominators is nonzero. As before, note that the numerator of the expression for $x$ differs from the denominator only in that each $a$ is replaced by the corresponding $d$; the numerator for $y$, in that each $b$ is replaced by the corresponding $d$; and the numerator for $z$, in that each $c$ is replaced by the corresponding $d$.

This third-order formula and the corresponding one for second-order determinants given earlier are called **Cramer's rule**. The theory of determinants shows that Cramer's rule works in exactly the same way for a system of $n$ linear equations in $n$ unknowns, where $n$ is any positive integer.

## Problems

1.  Evaluate the following determinants.

$$\begin{vmatrix} 1 & 2 \\ 3 & 4 \end{vmatrix}, \quad \begin{vmatrix} -2 & 7 \\ -5 & -6 \end{vmatrix}, \quad \begin{vmatrix} 2 & 0 & -1 \\ 3 & 2 & 6 \\ -4 & 5 & 0 \end{vmatrix}, \quad \begin{vmatrix} -2 & 4 & 6 \\ 3 & 0 & 1 \\ 1 & 1 & -7 \end{vmatrix}$$

2.  Solve the following systems of equations by using determinants.

$$6x + 7y = 18$$
$$9x - 2y = -48$$

$$2x + 3y + z = 4$$
$$x + 5y - 2z = -1$$
$$3x - 4y + 4z = -1$$

# 16

# Geometric Progressions & Series

A **geometric progression** is a sequence of numbers in which each term after the first is obtained from the one that precedes it by multiplying by a fixed number called the **ratio**:

$$a, ar, ar^2, ar^3, \ldots, ar^n, \ldots \quad (r = \text{the ratio})$$

For example,

| 2, | 4, | 8, | 16, | $\ldots$, | $r = 2$ |
|---|---|---|---|---|---|
| 2, | 6, | 18, | 54, | $\ldots$, | $r = 3$ |
| 1, | $\dfrac{1}{2}$, | $\dfrac{1}{4}$, | $\dfrac{1}{8}$, | $\ldots$, | $r = \dfrac{1}{2}$ |
| 2 | $-\dfrac{2}{3}$, | $\dfrac{2}{9}$, | $-\dfrac{2}{27}$, | $\ldots$, | $r = -\dfrac{1}{3}$ |

The sum $S$ of the first $n$ terms of a geometric progression is

$$S = a + ar + ar^2 + ar^3 + \cdots + ar^n$$

or

$$S = \frac{a(1 - r^{n+1})}{1 - r}$$

To prove that these two formulas are equivalent (assuming $r \neq 1$, since $r = 1$ makes the second formula meaningless), multiply the first formula through by $r$ to get

$$Sr = ar + ar^2 + ar^3 + \cdots + ar^{n+1}$$

and then subtract this result from the first formula, cancelling all the common terms:

$$S = a + ar + ar^2 + \cdots + ar^n$$

$$Sr = ar + ar^2 + ar^3 + \cdots + ar^{n+1}$$

From these cancellations, it's clear that

$$S - Sr = a - ar^{n+1}$$

or

$$S(1-r) = a(1 - r^{n+1})$$

which is equivalent to

$$S = \frac{a(1 - r^{n+1})}{1 - r}$$

To extend the sum $S$ to an infinite number of terms, use dots in the following way

$$a + ar + ar^2 + ar^3 + \cdots + ar^n + \cdots$$

This sum is called a **geometric series**. Its value—if it has one—is determined by examining the behavior of the **partial sum**

$$S = a + ar + ar^2 + ar^3 + \cdots + ar^n$$

as the positive integer $n$ increases. By using the formula for $S$ given above, this partial sum can be written in the form

$$S = \frac{a(1 - r^{n+1})}{1 - r} = \frac{a}{1 - r} - \frac{a}{1 - r}r^{n+1}$$

If the ratio $r$ is any number numerically less than 1, that is, if $|r| < 1$, then the second term on the right

$$\frac{a}{1 - r}r^{n+1}$$

approaches zero as $n$ increases. (As a number numerically less than one is raised to higher and higher powers, it gets smaller and smaller.)

This concept is expressed as

$$\frac{a}{1-r} r^{n+1} \to 0 \quad \text{as } n \to \infty$$

where the arrow means "approaches." For these values of $r$ we therefore have

$$S = a + ar + ar^2 + \cdots + ar^n \to \frac{a}{1-r} \quad \text{as } n \to \infty$$

which is what is meant by the statement that the formula

$$a + ar + ar^2 + \cdots + ar^n + \cdots = \frac{a}{1-r}$$

is valid for $|r| < 1$.

The formula for $a/(1 - r)$ can be used to show that any infinite repeating decimal represents a rational number. For example, to see why

$$0.3333\ldots = \frac{1}{3}$$

is true, use the meaning of the decimal and apply the $a/(1 - r)$ formula:

$$0.3333\ldots = \frac{3}{10} + \frac{3}{10^2} + \frac{3}{10^3} + \cdots$$

$$= \frac{3}{10} + \left(\frac{3}{10}\right)\left(\frac{1}{10}\right) + \left(\frac{3}{10}\right)\left(\frac{1}{10}\right)^2 + \cdots$$

$$= \frac{\dfrac{3}{10}}{1 - \dfrac{1}{10}} = \frac{3}{10} \cdot \frac{10}{9} = \frac{1}{3}$$

# Problems

1. If a pond has 5 lily pads, and if each grows into two lily pads every twelve hours and none die, how many will there be a week later? Use geometric series and the fact that after one day there will be $5 \cdot 2 \cdot 2 = 5 \cdot 4$.

2. A ball is dropped from the height of 81 inches. If it rebounds 2/3 of the distance it falls, use geometric series to find the total distance it has traveled if it is caught at the top of the fourth bounce.

3. In the preceding problem, what's the total distance traveled by the ball before it comes to rest?

4. In an infinite sequence of nested equilateral triangles, the vertices of each triangle after the first are the midpoints of the sides of the preceding triangle. Use geometric series to find the sum of the perimeters of all the triangles if the perimeter of the first triangle is 12.

5. Sum the following geometric series.

   $1 + \dfrac{1}{2} + \dfrac{1}{4} + \cdots$

   $4 - 2 + 1 - \cdots$

   $9 + 6 + 4 + \cdots$

   $6 - 2 + \dfrac{2}{3} - \cdots$

   $3 + \sqrt{3} + 1 + \cdots$

   $\sqrt{12} + \sqrt{6} + \sqrt{3} + \cdots$

6. Express $0.777\ldots$, $0.343434\ldots$, and $3.72444\ldots$ as fractions by using geometric series.

# 17

# Arithmetic Progressions

An **arithmetic progression** is a sequence of numbers in which each term after the first is obtained from the one that precedes it by adding a fixed number called the **common difference**:

$$a, a + d, a + 2d, \ldots, a + (n - 1)d$$

The simplest and most important arithmetic progression is the first $n$ positive integers:

$$1, 2, 3, \ldots, n$$

If $S$ denotes the sum of this progression,

$$S = 1 + 2 + 3 + \cdots + n$$

then a formula for $S$ has many uses. To find this formula, write the sum twice, once as given and the second time in reverse order:

$$
\begin{array}{ccccccccc}
S &=& 1 &+& 2 &+& 3 &+& \cdots &+& n \\
S &=& n &+& (n-1) &+& (n-2) &+& \cdots &+& 1
\end{array}
$$

Adding the two expressions yields $2S$ on the left; and because each column on the right adds up to $n + 1$ and there are $n$ columns, we get

$$2S = n(n + 1)$$

or

$$S = \frac{n(n+1)}{2}$$

## Problems

1. Use an arithmetic progression to find a formula for the sum of the first $n$ odd positive integers.

# 18

# Permutation & Combinations

The techniques of counting covered in this chapter are useful in a variety of situations.

In the notation of counting, $n$ is a positive integer and the product of all positive integers up to $n$ is $1 \cdot 2 \cdot 3 \cdots n$. This product is denoted by $n!$ and called "$n$ factorial":

$$n! = 1 \cdot 2 \cdot 3 \cdots n$$

Thus,

$$1! = 1$$
$$2! = 1 \cdot 2 = 2$$
$$3! = 1 \cdot 2 \cdot 3 = 6$$
$$4! = 1 \cdot 2 \cdot 3 \cdot 4 = 24$$

and so on. For technical reasons, $0!$ is defined to be 1. A few calculations show that factorials increase rapidly:

| | | |
|---|---|---|
| $5! = 120$ | $6! = 720$ | $7! = 5040$ |
| $8! = 40,320$ | $9! = 362,880$ | $10! = 3,628,800$ |

To absorb the reasoning behind counting, consider a simple example: a trip from city $A$ through city $B$ to city $C$. Suppose that it's possible to travel from $A$ to $B$ by 4 different routes and from $B$ to $C$ by 3 different routes. Then the total number of different routes from $A$ through $B$ to $C$ is $4 \cdot 3 = 12$, because we can go from $A$ to $B$ in any one of 4 ways, and for each of these ways there are 3 ways of going on from $B$ to $C$.

The basic principle is: if two successive independent decisions are to be made, and if there are $c_1$ choices for the first and $c_2$ choices for the second, then the total number of ways of making these two decisions is the product $c_1 c_2$. This same principle is valid for any number of decisions.

By applying this principle, we can answer the question: in how many ways can we arrange $n$ distinct objects in order, that is, with a first, a second, a third, and so on? There are $n$ choices for the first object. After the first object is chosen, there are $n - 1$ choices for the second, then $n - 2$ choices for the third, and so on. By the basic principle stated above, the total number of ordering is therefore

$$n(n - 1)(n - 2) \cdots 3 \cdot 2 \cdot 1 = n!$$

Each ordering of a set of objects is called a **permutation** of those objects, therefore

The number of permutations of $n$ objects is $n!$

There are $5! = 120$ ways of arranging 5 books on a shelf, for example, and there are $9! = 362{,}880$ batting orders for a 9-player baseball team.

Suppose again that we have $n$ distinct objects. A more-general question is: in how many ways can $k$ of them be chosen in order? Each such ordering is called a permutation of $n$ objects taken $k$ at a time, and the total number of these permutations is denoted by $P(n, k)$. There are $n$ choices for the first, $n - 1$ choices for the second, $n - 2$ choices for the third, and $n - (k - 1) = n - k + 1$ choices for the $k$th. The total number of these permutations is therefore

$$P(n, k) = n(n - 1)(n - 2) \cdots (n - k + 1)$$

By inserting additional factors and cancelling them out again, we can write $P(n, k)$ by using factorials. The number of permutations of $n$ objects taken $k$ at a time is

$$P(n,k) = n(n-1)(n-2)\cdots(n-k+1) = \frac{n!}{(n-k)!}$$

For example, if we have 6 books but only 3 available spaces on a bookshelf, then the number of ways of filling these spaces with the books on hand (counting the order of books as they stand on the shelf) is

$$P(6,3) = \frac{6!}{(6-3)!} = \frac{6!}{3!} = 6 \cdot 5 \cdot 4 = 120$$

The number of ways (counting the order of the cards) in which a 5-card poker hand can be dealt from a deck of 52 cards is

$$P(52,5) = \frac{52!}{(52-5)!} = \frac{52!}{47!} = 52 \cdot 51 \cdot 50 \cdot 49 \cdot 48 = 311,875,200$$

The order of cards in a poker hand is immaterial to the value of the hand, however, so the number of distinct poker hands is a much smaller number, which is taken into account by the next topic, combinations.

A set of $k$ objects chosen from a given set of $n$ objects, without regard to the order in which they're arranged, is called a **combination** of $n$ objects taken $k$ at a time. The total number of such combinations is denoted by $C(n, k)$ or, more frequently, by $\binom{n}{k}$. For reasons explained in Chapter 19, the numbers $\binom{n}{k}$ are called **binomial coefficients**.

Permutations and combinations are related in a simple way. Each permutation of $n$ objects taken $k$ at a time consists of a choice of $k$ objects (a combination) followed by an ordering of these $k$ objects. But there are $\binom{n}{k}$ ways to choose $k$ objects, and then $k!$ ways to order them, so

$$P(n,k) = \binom{n}{k} \cdot k! \quad \text{or} \quad \binom{n}{k} = \frac{P(n,k)}{k!}$$

So the total number of combinations of $n$ objects taken $k$ at a time is

$$\binom{n}{k} = \frac{n!}{k!(n-k)!}$$

For example, the number of committees of 4 people that can be chosen from a group of 7 people is

$$\binom{7}{4} = \frac{7!}{4!3!} = \frac{7 \cdot 6 \cdot 5}{2 \cdot 3} = 35$$

The number of different 5-card poker hands that can be dealt from a deck of 52 cards is

$$\binom{52}{5} = \frac{52!}{5!47!} = \frac{52 \cdot 51 \cdot 50 \cdot 49 \cdot 48}{2 \cdot 3 \cdot 4 \cdot 5} = 2,598,960$$

## Problems

1. Compute the following values.

$$\frac{9!}{6!} \quad \frac{13!}{7!} \quad \binom{18}{3} \quad \binom{36}{4}$$

2. How many different orders of finishing are there in a race of 8 runners? How many different possibilities are there for the first three places (1st, 2nd, and 3rd)?

3. A club has 14 members. In how many ways can a president, vice president, and secretary be chosen?

4. A student can choose to answer any 3 out of 9 questions on a test. In how many ways can the student choose the questions?

5. In how many ways can 6 fish be chosen from an aquarium containing 15 fish?

# 19       The Binomial Theorem

The binomial theorem is a general formula for the expanded $n$-factor product

$$(a + b)^n = (a + b)(a + b) \cdots (a + b)$$

Computing the first few cases by repeated brute-force multiplications yields

$$(a + b)^1 = a + b$$
$$(a + b)^2 = a^2 + 2ab + b^2$$
$$(a + b)^3 = a^3 + 3a^2b + 3ab^2 + b^3$$
$$(a + b)^4 = a^4 + 4a^3b + 6a^2b^2 + 4ab^3 + b^4$$
$$(a + b)^5 = a^5 + 5a^4b + 10a^3b^2 + 10a^2b^3 + 5ab^4 + b^5$$

It's clear that the expansion of $(a + b)^n$ begins with $a^n$ and ends with $b^n$, and also the intermediate terms involve steadily decreasing powers of $a$ and steadily increasing powers of $b$ so that the sum of each of the two exponents is exactly $n$ in each term. What's unclear is the way that the coefficients are calculated.

The general form of the expansion of $(a + b)^n$—the **binomial theorem**—is

$$(a+b)^n = a^n + na^{n-1}b + \frac{n(n-1)}{2}a^{n-2}b^2 + \frac{n(n-1)(n-2)}{2 \cdot 3}a^{n-3}b^3 + \cdots$$

$$+ \frac{n(n-1)(n-2)\cdots(n-k+1)}{1 \cdot 2 \cdot 3 \cdots k}a^{n-k}b^k + \cdots + b^n$$

To prove the binomial theorem, it's necessary to understand how the coefficients are formed. Observe that each term of the expansion of $(a + b)^n$ is the product of $n$ letters, one taken from each factor of the product

$$(a + b)^n = (a + b)(a + b) \cdots (a + b), \quad n \text{ factors}$$

Thus, $a^{n-k}b^k$ is the product of $k$ $b$'s and the rest $a$'s. The number of ways that this can be done is $\binom{n}{k}$, the number of combinations of $n$ objects taken $k$ at a time, so this is the number of times the term $a^{n-k}b^k$ occurs in the expansion. The coefficient of $a^{n-k}b^k$ on the right side of the binomial theorem must therefore be $\binom{n}{k}$, and because

$$\binom{n}{k} = \frac{n!}{k!(n-k)!} = \frac{n(n-1)(n-2)\cdots(n-k+1)}{1 \cdot 2 \cdot 3 \cdots k}$$

the proof is complete.

Using the binomial theorem to find $(a + b)^5$, for example, yields

$$(a+b)^5 = a^5 + \binom{5}{1}a^4b + \binom{5}{2}a^3b^2 + \binom{5}{3}a^2b^3 + \binom{5}{4}ab^4 + b^5$$

$$= a^5 + \frac{5!}{1!4!}a^4b + \frac{5!}{2!3!}a^3b^2 + \frac{5!}{3!2!}a^2b^3 + \frac{5!}{4!1!}ab^4 + b^5$$

$$= a^5 + 5a^4b + \frac{5 \cdot 4}{2}a^3b^2 + \frac{5 \cdot 4 \cdot 3}{2 \cdot 3}a^2b^3 + \frac{5 \cdot 4 \cdot 3 \cdot 2}{2 \cdot 3 \cdot 4}ab^4 + b^5$$

$$= a^5 + 5a^4b + 10a^3b^2 + 10a^2b^3 + 5ab^4 + b^5$$

## Problems

1. Use the binomial theorem to expand the following expressions.

$(a+b)^9$

$(2a+b)^6$

$(2a-3b)^5$

# 20      Mathematical Induction

In Chapter 17, we derived the formula for the sum of the first $n$ positive integers:

$$1 + 2 + 3 + \cdots + n = \frac{n(n+1)}{2}$$

It's also sometimes useful to know the corresponding formula for the sum of the first $n$ squares:

$$1^2 + 2^2 + 3^2 + \cdots + n^2 = \frac{n(n+1)(2n+1)}{6}$$

Though we can derive this formula by using several methods (omitted here), we can't use the same method from Chapter 17 because the terms on the left don't form an arithmetic progression. One method of proof that *does* work, however, is the

**Principle of Mathematical Induction.** A statement $S(n)$ that depends on a positive integer $n$ is true for all $n$ if it satisfies the following two conditions:

1.  $S(n)$ is true for $n = 1$.

2.  If $S(n)$ is true for $n = k$, then it is also true for $n = k + 1$.

To use mathematical induction to prove the formula for sum of the first $n$ squares given above, let

$$S(n) = 1^2 + 2^2 + 3^2 + \cdots + n^2 = \frac{n(n+1)(2n+1)}{6}$$

First, verify that $S(n)$ is true for $n = 1$:

$$1^2 = \frac{1 \cdot 2 \cdot 3}{6}$$

Next, assume that $S(n)$ is true for $n = k$:

$$1^2 + 2^2 + 3^2 + \cdots + k^2 = \frac{k(k+1)(2k+1)}{6}$$

and then try to prove $S(n)$ for $n = k + 1$:

$$1^2 + 2^2 + 3^2 + \cdots + k^2 + (k+1)^2 = \frac{(k+1)(k+2)(2k+3)}{6}$$

Start with the left side and use the induction hypothesis to obtain the right side:

$$1^2 + 2^2 + 3^2 + \cdots + k^2 + (k+1)^2$$

$$= \frac{k(k+1)(2k+1)}{6} + (k+1)^2$$

$$= (k+1)\left[\frac{k(2k+1)}{6} + k + 1\right]$$

$$= \frac{(k+1)(2k^2 + 7k + 6)}{6}$$

$$= \frac{(k+1)(k+2)(2k+3)}{6}$$

Having proved steps 1 and 2, we conclude by the principle that $S(n)$ is true for all $n$.

## Problems

1.  Use mathematical induction to prove that $1 + 3 + 5 + \ldots + (2n - 1) = n^2$ is true for all positive integers $n$.

# 21 Solutions

## Chapter 1

1. $(3a - b) - [2a - (a + b)] = (3a - b) - [2a - a - b]$

$$= 3a - b - 2a + a + b$$

$$= 2a$$

$[(a + 3b) - a] - [a - (a - 3b)] = [a + 3b - a] - [a - a + 3b]$

$$= a + 3b - a - a + a - 3b]$$

$$= 0$$

$a - \{2a - [b - (3a - 2b)]\} = a - \{2a - [b - 3a + 2b]\}$

$$= a - \{2a - b + 3a - 2b\}$$

$$= a - 2a + b - 3a + 2b$$

$$= 3b - 4a$$

2. $\dfrac{2}{5} \cdot \dfrac{m+3}{7} + \dfrac{1}{2} = \dfrac{2m+6}{35} + \dfrac{1}{2} = \dfrac{2m+6}{35} \cdot \dfrac{2}{2} + \dfrac{1}{2} \cdot \dfrac{35}{35}$

$$= \dfrac{4m + 12 + 35}{70} = \dfrac{4m + 47}{70}$$

3. $6(2x - 3y + 5)$      $4x^2(2 - 3xy - 7x^2 z)$      $3abc(3 + abc)$

4. $\dfrac{a}{b} - \dfrac{b}{a} = \dfrac{a^2}{ab} - \dfrac{b^2}{ab} = \dfrac{a^2 - b^2}{ab}$

$\dfrac{1}{1 + \dfrac{1}{x-1}} = \dfrac{1}{\dfrac{x-1}{x-1} + \dfrac{1}{x-1}} = \dfrac{1}{\dfrac{x}{x-1}} = \dfrac{x-1}{x}$

$\dfrac{1}{x-y}\left(\dfrac{x}{y} - \dfrac{y}{x}\right) = \dfrac{1}{x-y}\left(\dfrac{x^2 - y^2}{xy}\right) = \dfrac{1}{x-y}\left(\dfrac{(x+y)(x-y)}{xy}\right) = \dfrac{x+y}{xy}$

$\dfrac{(x+a)^2 - x^2}{a} = \dfrac{x^2 + 2xa + a^2 - x^2}{a} = \dfrac{2xa + a^2}{a} = 2x + a$

5. $\dfrac{x-2}{y} \bigg/ \dfrac{z}{x+2} = \dfrac{x-2}{y} \cdot \dfrac{x+2}{z} = \dfrac{x^2 - 4}{yz}$

## Chapter 2

1. $x^5(x^2)^3 = x^5 x^6 = x^{11}$

$y^4(y^2(y^5)^2)^3 = y^4(y^2 y^{10})^3 = y^4(y^{12})^3 = y^4 y^{36} = y^{40}$

$t^4(t^3(t^{-2})^5)^4 = t^4(t^3 t^{-10})^4 = t^4(t^{-7})^4 = t^4 t^{-28}$

$\dfrac{(x^{-2})^3 y^8}{x^{-5}(y^4)^{-3}} = \dfrac{x^{-6} y^8}{x^{-5} y^{-12}} = \dfrac{y^{8+12}}{x^{6-5}} = \dfrac{y^{20}}{x}$

$\dfrac{(x^2 y^4)^3}{(x^5 y^2)^{-4}} = \dfrac{x^6 y^{12}}{x^{-20} y^{-8}} = x^{6+20} y^{12+8} = x^{26} y^{20}$

$\left(\dfrac{(x^2 y^{-5})^{-4}}{(x^5 y^{-2})^{-3}}\right)^2 = \dfrac{(x^2 y^{-5})^{-8}}{(x^5 y^{-2})^{-6}} = \dfrac{x^{-16} y^{40}}{x^{-30} y^{12}} = x^{30-16} y^{40-12} = x^{14} y^{28}$

2. $16000 = 16 \cdot 1000 = 2^4 \cdot 10^3 = 2^4 \cdot (2 \cdot 5)^3 = 2^4 \cdot 2^3 \cdot 5^3 = 2^7 \cdot 5^3$, therefore $m = 7$ and $n = 3$.

3. $8^{1000}/2^5 = (2^3)^{1000}/2^5 = 2^{3000}/2^5 = 2^{2995}$

4. $a^n b^{4n}$     $12a^2b$     $x^{10}y^{10}$

   $b^2 + a^2$     $\dfrac{(x+y)^2}{xy}$     $1$

5. $12$    $0.8$    $\dfrac{3}{4}$    $-\dfrac{1}{3}$

   $-10$    $5\sqrt{5}$    $5$    $3\sqrt{2}$

   $2\sqrt{3}$    $3\sqrt{2}$    $2\sqrt{3}$    $8\sqrt[3]{2}$

   $\sqrt{2a}$    $ab^2$    $a\sqrt[4]{a}$    $1/2$

6. $\dfrac{30}{\sqrt{6}} = \dfrac{30}{\sqrt{6}} \cdot \dfrac{\sqrt{6}}{\sqrt{6}} = \dfrac{30\sqrt{6}}{6} = 5\sqrt{6}$

   $\dfrac{\sqrt{6}+2}{\sqrt{6}-2} = \dfrac{\sqrt{6}+2}{\sqrt{6}-2} \cdot \dfrac{\sqrt{6}+2}{\sqrt{6}+2} = \dfrac{4\sqrt{6}+10}{(\sqrt{6})^2 - 2^2} = 5 + 2\sqrt{6}$

   $\dfrac{2}{\sqrt{7}+\sqrt{5}} = \dfrac{2}{\sqrt{7}+\sqrt{5}} \cdot \dfrac{\sqrt{7}-\sqrt{5}}{\sqrt{7}-\sqrt{5}} = \dfrac{2(\sqrt{7}-\sqrt{5})}{(\sqrt{7})^2 - (\sqrt{5})^2} = \sqrt{7} - \sqrt{5}$

7. $25^{3/2} = (25^{1/2})^3 = 5^3 = 125$

   $32^{3/5} = (32^{1/5})^3 = 2^3 = 8$

   $32^{-4/5} = (32^{1/5})^{-4} = 2^{-4} = \dfrac{1}{2^4} = \dfrac{1}{16}$

   $(-8)^{7/3} = ((-8)^{1/3})^7 = (-2)^7 = -128$

8. $\dfrac{5a^3}{b}$    $16a^2b$    $ab$

$a-b$    $a^3$    $\dfrac{3b^2c^2}{4a^2}$

## Chapter 3

1.  $x^7 + 2x^6 - 8x^5 - 2x^4 + x^3 + 2x^2 + x - 8$

$3x^5 - 2x^3 - 11x$

$6x^5 + 5x^4 - 24x^3 - 39x^2 + 8x + 36$

$2x^7 - 8x^6 + 16x^4 - 8x^3 + 6x^2 - 24x + 12$

$x^4 - 1$

## Chapter 4

1.  $(x-3)(x+2)$          $(x-2)^2$          $x(x+6)^2$

$(x^2+4)(x+2)(x-2)$    $x(x-4)(x+1)$    $(2x+4)(2x-3)$

$(2x-4)(5x+2)$

2.  $x^3 - 27 = (x-3)(x^2+3x+9)$

$8x^3 - 125 = (2x-5)(4x^2+10x+25)$

3.  $x^3 + 64 = (x+4)(x^2-4x+16)$

$27x^3 + 8 = (3x+2)(9x^2-6x+4)$

4.  4 and –7

11 and –3

5/2 and –3

7/3 and –3/2

# Chapter 5

1. $\dfrac{9 \pm \sqrt{21}}{10}$    $\dfrac{-7 \pm \sqrt{13}}{6}$

   $\dfrac{3 \pm 2\sqrt{-2}}{17}$    $\dfrac{-1 \pm \sqrt{-3}}{2}$

# Chapter 6

1. $x < -6$

   $x > 3$

2. $x = \pm 2$    $x = \pm 3$    $x = \pm 6$    $x = -1, 5$    $x = -4, -2$

3. $x > 1$ or $x < 0$

   $x > 3$ or $x < -5$

# Chapter 7

1. 12, 5, 13

2. 35

3. (4, 5)

# Chapter 8

1. $f(0) = -3, f(1) = 1, f(2) = 5, f(3) = 9$

2. $g(0) = -4, g(1) = -\frac{1}{2}, g(-\frac{1}{2}) = -20/7$

3. $h(x^3) = x^9 - 3x^6 + 5x^3 - 1$

4. $F[F(x)] = x$

5. $A = s^2$
   $A = p^2/16$

6. $A = c^2/4\pi$

7. $h = \frac{1}{2}\sqrt{3}\, b$

8. $d = 5t$

# Chapter 9

1.

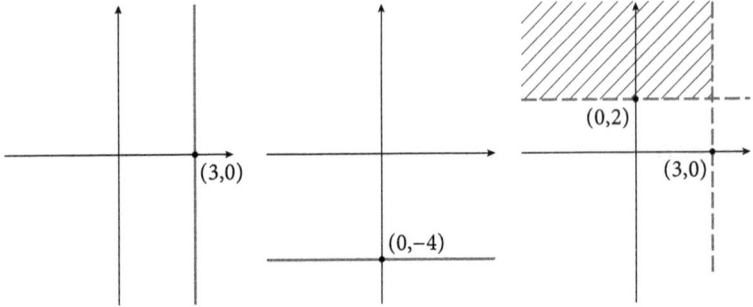

2.

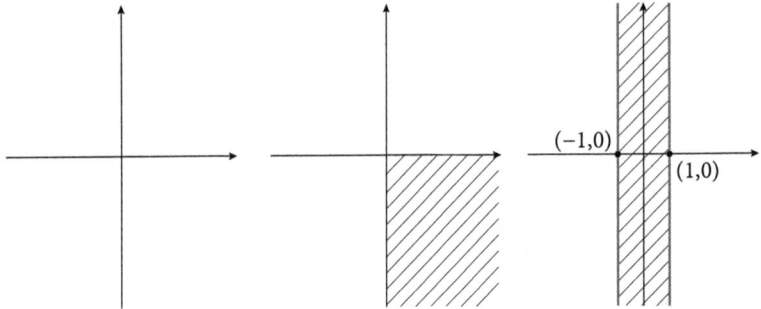

3. 2; 0; 1

4. The slopes of the lines (−3) are equal, hence they are parallel.

5. The parallel line is $(y + 3)/(x + 2) = -\frac{1}{2}$ or $x + 2y = -8$.

   The perpendicular line is $(y + 3)/(x + 2) = 2$ or $y = 2x + 1$.

# Chapter 10

1. $x^2 + y^2 = 4$ $\qquad\qquad (x+2)^2 + y^2 = 49$

   $(x-3)^2 + (y-6)^2 = \dfrac{1}{4}$ $\quad (x-1)^2 + (y-2)^2 = 25$

2. The points of intersection are (–2, 2) and (5, 3).

3. (–3, 6), 3

   (4, 0), 2

   (0, –2), 1

   (–3, –1), 2

   (8, –7), 4

# Chapter 11

1. $\left(0, \dfrac{1}{8}\right), y = -\dfrac{1}{8}$ $\quad \left(0, 2\right), y = -2$

   $\left(0, -\dfrac{1}{20}\right), y = \dfrac{1}{20}$ $\quad \left(0, -3\right), y = 3$

2. $x^2 = 12y \quad x^2 = 64y \quad x^2 = -4y \quad 5x^2 = -2y$

3. $(2, -3), \left(2, -2\dfrac{3}{4}\right)$, up $\qquad (3, -25), \left(3, -24\dfrac{7}{8}\right)$, up

   $(-2, 9), \left(-2, 8\dfrac{3}{4}\right)$, down $\quad (-2, 6), \left(-2, 5\dfrac{1}{2}\right)$, down

# Chapter 12

1. Imaginary       Real and distinct

   Real and distinct      Real and equal

2. $\text{sum} = \dfrac{7}{4}, \text{product} = -\dfrac{13}{4}$

   imaginary roots

   $\text{sum} = -\dfrac{1}{2}, \text{product} = -1$

3. $x^2 - 5x - 24 = 0$

   $x^2 - 4x - 1 = 0$

   $15x^2 - x - 6 = 0$

4.

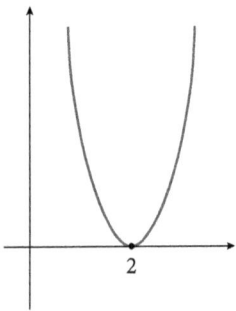

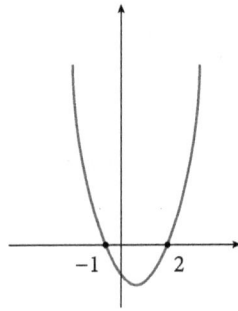

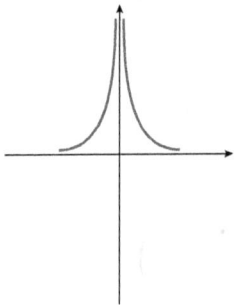

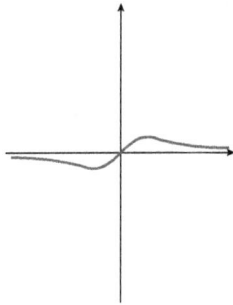

# Chapter 13

1. $2 = \log_5 25$    $5 = \log_2 32$    $-2 = \log_5 \dfrac{1}{25}$    $0.5 = \log_{81} 9$

   $0 = \log_7 1$    $-1 = \log_{10} \dfrac{1}{10}$    $\dfrac{4}{5} = \log_{32} 16$    $0.75 = \log_{16} 8$

2. $10^1 = 10$    $4^{3/2} = 8$    $2^3 = 8$    $5^{-3} = \dfrac{1}{125}$

   $10^{-2} = 0.01$    $7^3 = 343$    $5^3 = 125$    $10^{-1} = 0.1$

3. $1$    $4$    $4$    $\dfrac{3}{2}$

   $\dfrac{2}{3}$    $0$    $\dfrac{7}{9}$    $\dfrac{7}{6}$

4. $2187$    $243$    $256$    $16$

5. $243$    $\dfrac{1}{81}$    $7$    $36$

# Chapter 14

1. $Q(x) = x - 2; R(x) = 0$

   $Q(x) = x^2 - x - 6; R(x) = 0$

   $Q(x) = 3x^2 - 2x + 1; R(x) = -x + 7$

   $Q(x) = x^3 - 2x^2 + 5; R(x) = 0$

2. $(x-1)(x^2 + 1)$

   $(x+2)(x+6)(x-3)$

   $(x-3)(x+1)^2$

3. The other factor is $x^{n-1} + x^{n-2} + x^{n-3} + \cdots + x + 1$.

# Chapter 15

1. $-2, 47, -83, 108$

2. $x = -4, y = 6$
   $x = -3, y = 2, z = 4$

# Chapter 16

1. $5 \cdot 4^7 = 81920$

2. 4 down trips $+$ 4 up trips $= \dfrac{81(1-\frac{2}{3}^{3+1})}{1-\frac{2}{3}} + \dfrac{54(1-\frac{2}{3}^{3+1})}{1-\frac{2}{3}}$

   $= 195 + 130 = 325$

3. All down trips $+$ All up trips $= \dfrac{81}{1-\frac{2}{3}} + \dfrac{54}{1-\frac{2}{3}} = 243 + 162 = 405$

4. $\dfrac{a}{1-r} = \dfrac{12}{1-\frac{1}{2}} = 24$

5. $\dfrac{a}{1-r} = \dfrac{1}{1-(1/2)} = 2$

   $\dfrac{a}{1-r} = \dfrac{4}{1-(-1/2)} = 8/3$

   $\dfrac{a}{1-r} = \dfrac{9}{1-(2/3)} = 27$

   $\dfrac{a}{1-r} = \dfrac{6}{1-(-1/3)} = \dfrac{9}{2}$

   $\dfrac{a}{1-r} = \dfrac{3}{1-(1/\sqrt{3})} = \dfrac{3\sqrt{3}(\sqrt{3}+1)}{2}$

   $\dfrac{a}{1-r} = \dfrac{\sqrt{12}}{1-(1/\sqrt{2})} = 2\sqrt{6}(\sqrt{2}+1)$

6. $7/9, 34/99, 838/225$

# Chapter 17

1. $1 + 3 + 5 + \ldots + (2n - 1) = n^2$

# Chapter 18

1. 504, 1235520, 816, 58905

2. $8! = 40,320$
   $8 \cdot 7 \cdot 6 = 336$

3. $14 \cdot 13 \cdot 12 = 2184$

4. $\binom{9}{3} = 84$

5. $\binom{15}{6} = 5005$

# Chapter 19

1. $a^9 + 9a^8b + 36a^7b^2 + 84a^6b^3 + 126a^5b^4 + 126a^4b^5 + 84a^3b^6 +$

   $36a^2b^7 + 9ab^8 + b^9$

   $64a^6 + 192a^5b + 240a^4b^2 + 160a^3b^3 + 60a^2b^4 + 12ab^5 + b^6$

   $32a^5 - 240a^4b + 720a^3b^2 - 1080a^2b^3 + 810ab^4 - 243b^5$

# Chapter 20

1. Let

$$S(n) = 1 + 3 + 5 + \cdots + (2n-1) = n^2$$

First, verify that $S(n)$ is true for $n = 1$:

$$1 = 1^2$$

Next, assume that $S(n)$ is true for $n = k$:

$$1 + 3 + 5 + \cdots + (2k-1) = k^2$$

and then try to prove $S(n)$ for $n = k + 1$:

$$1 + 3 + 5 + \cdots + (2k-1) + [2(k+1)-1] = (k+1)^2$$

Start with the left side and use the induction hypothesis to obtain the right side:

$$1 + 3 + 5 + \cdots + (2k-1) + [2(k+1)-1]$$

$$= [1 + 3 + 5 + \cdots + (2k-1)] + [2(k+1)-1]$$

$$= k^2 + [2(k+1)-1]$$

$$= k^2 + [2k+2-1]$$

$$= k^2 + 2k + 1$$

$$= (k+1)^2$$

Having proved steps 1 and 2, we conclude by the principle that $S(n)$ is true for all $n$.

# Index